CELEBRATED AMERICAN CAVES

CELEBRATED AMERICAN CAVES

Edited by
CHARLES E. MOHR AND HOWARD N. SLOANE

With a Foreword by Alexander Wetmore

Contributors

BLACK — BRIDGES — BRUCKER — DAVIES
FOOTE — HALLIDAY — HARRINGTON — HIBBEN
JACKSON — LIX — MOHR — NICHOLAS
SAWTELLE — SLOANE — WHITE

RUTGERS   UNIVERSITY   PRESS
New Brunswick   *New Jersey*

Copyright © 1955
By the Trustees of Rutgers College in New Jersey
Library of Congress Catalogue Number: 55-12228
Fifth Printing

Appreciation is gratefully expressed herewith to the following for their courtesy in allowing the reprinting of copyrighted material listed below:

To Howard Zahniser, editor of *The Living Wilderness,* for "Mammoth Cave's Underground Wilderness" by Henry W. Lix, copyright 1946.

To Edward M. Weyer, editor of *Natural History,* and to Alonzo M. Pond for passages from "Lost John of Mummy Ledge," copyright 1937, and to M. R. Harrington for "Idol of the Cave," copyright 1951.

To the Viking Press, and to Ivan T. Sanderson for passages from *Caribbean Treasure,* copyright 1939.

To the National Speleological Society for "The Devil's Sinkhole" and portions of "Bat Blitz" by Patrick J. White, for parts of "Texas Caves Served in Three Wars" by Charles E. Mohr, and for parts of "Falcons Prey on Ney Cave Bats," by Kenneth E. Stager (in part, originally published in *The Condor),* all from Bulletin Ten, *Caves of Texas,* copyright 1948; and for the use of photographs by Russell T. Neville.

To The Potomac Appalachian Trail Club and to Tom Culverwell for illustrations and maps from "Mountaineering Under West Virginia" from the Club's *Bulletin,* copyright 1941.

Manufactured in the United States of America

Foreword

Seventy years have passed since the appearance of Horace C. Hovey's work, *Celebrated American Caverns*, so that the present volume, sponsored by the National Speleological Society, is timely in view of the steadily growing attraction of this field. Most of the exposed surface of land and sea has been examined, at least from the air, but there remains the subterranean world, a major part of whose secrets have not been penetrated. The exploration of caverns, ages-old, dark passages cut through sub-surface rock by the slow processes of natural erosion, still offers the challenge of novelty, as well as the possibility of increasing human knowledge—of details of size and extent, of strange and unusual secondary mineral deposits, of traces of ancient human habitation, and of curious animals of specialized form. Add to this the adventure of entrance into places unknown to other men, and you have the magnet that has drawn together in cooperative association the spelunkers and the speleologists—the amateurs and the professionals—who form the steadily growing membership of our society. Founded in 1941 by a small group of enthusiasts, the National Speleological Society is now an active and growing body whose members are spread widely through North America and are scattered as well through other parts of the world. The aim of the organization is to foster interest in the knowledge to be gained from cave exploration, and to protect from thoughtless vandalism the natural features of the underground world. This book has been prepared with these points in mind.

Caves have served as one form of shelter for some of the earliest races of man, traces of whose occupancy may still remain—charcoal of ancient fires, broken bones of animal food, primitive implements of various kinds—tantalizing suggestions of phases of human history of which we may never have further information. In the Old

World, occupied caverns are sometimes of great antiquity; one recent exploration by Dr. Ralph Solecki in Shanidar Cave in the Zagros Mountains of northern Iraq suggested that there had been occupancy as long as 75,000 years ago. And, it may be added, Shanidar still serves as the winter home of families of Kurdish herders.

In America there is nothing in the human field to approach such age, but finds of animal bones may date from the years of the Pleistocene, or even the still earlier Pliocene, in which man is not known. Rarely, the huge bones of elephant-sized ground sloths have been found, but more often the remains are fragments of no apparent interest whatever to the layman. To the trained eyes of scientists, though, they may be of outstanding value in giving information on some strange creature of a kind no longer living. A single tooth or a broken bit of bone may serve as another of the little windows which give some additional view of the life of the world before the time of man.

In the present volume, competent students have brought together records of caves of wide geographic distribution and of much variety in kind, in accounts that are certain to arouse interest in caverns and their exploration. Which leads naturally to earnest entreaty to those who enter caves through casual opportunity always to use care not to mar those natural features that may be attractive to others. Above all, should human or other animal remains be seen, leave them undisturbed and bring them to the attention of professionals, so that the addition to knowledge that they may afford will not be lost.

ALEXANDER WETMORE

Editors' Note

What is a cave? A natural shelter for early man and a refuge from savage beasts? A storehouse for prehistoric relics; a hidden tomb? A laboratory for the evolutionist; a haven from bombs? A temple; a hide-out for murderers? A background for legends; a last frontier to be sought out and conquered?

A cave may have many uses and mean different things to different people—more than could be told in a single volume. But certain caves have special significance. The ones included here have each some paramount importance, whether it be in the field of exploration, science, history, or legend. Some of these caves are world famous. The names of some are virtually household words, for a few have been visited by several million persons; but there are many caves whose stories are known only to a few scientists or explorers.

It is our privilege to present both those long celebrated and those which deserve to be. To everyone who reads these stories the underworld is sure to take on new and adventurous meaning. And those who descend into the earth in the endless variety of commercial caves, or who probe into the infinite number of "wild" ones, can best share the excitement of the explorer, the irresistible urge to know what lies in the darkness beyond.

Cave exploring has come of age in the last few decades, with the acceptance of safe standards for equipment and conduct; recognition of the scientific importance and irreplaceable nature of the fossils, the artifacts, and the wildlife of caves; and respect for geological wonders that have been ages in the making.

Throughout the world cave explorers have banded together, dedicated to the exploration and protection of all that lies within this strange underworld. In the United States it is the National Speleological Society, Inc., that has marshaled the efforts of scientists and explorers, photographers and writers, in a widespread program of study and exploration.

While it is true that much of the activity described in this book preceded the establishment of the National Speleological Society in 1941, the greatest portion was carried on by individuals who became its first members, and their findings might remain unknown except for the research and explorations of the society's many members.

For it is people who make the world of caves so fascinating: not only the myriad bats that live in the caves, but also the people who have encountered them in their overpowering flights, people who have studied them and banded them; not solely the enigma of the prehistoric cave dwellers, but the patient sleuthing of modern students of the past; not alone the slow evolution of strange creatures, the almost timeless growth of caves, or the endless mystery of their labyrinthine extent, but the burning unrest of the explorer, the thrill of adventure and discovery, and the bitterness of defeat, even the final defeat—death, where discovery beckoned.

This is the story of American caves, the first since Horace C. Hovey's *Celebrated American Caverns,* which appeared originally in 1882. Hovey included much about caves in Europe and in other parts of the world. Since Hovey's time, many books have been written about European caves, but nothing comprehensive about those of the Western Hemisphere. The telling of the story of American speleology is long overdue.

The editors owe a debt of gratitude to the thirteen authors with whom they are here associated, and who have unselfishly supported this new attempt to tell the story of American caves.

Words alone, however, cannot give a complete picture of this strange underground world. Small wonder that the photographers among cavers have used all the technical skill at their command to capture a visual record. With ingenuity and perseverance, they have to a great extent overcome problems of difficult terrain, imperfect lighting, pervasive dust and grit, mud and water, high humidity, and many other, often unexpected, elements that hamper the cave photographer and are hard on film and cameras. We are presenting here what we believe to be the largest and most varied collection of cave pictures ever published in the Western Hemisphere. Tom Culverwell has kindly allowed the use of his unique drawings to help convey the special quality of difficult or dangerous cave exploration.

The editors acknowledge also several kinds of special assistance.

Invaluable was that of many of the society's officers, members, and chapters; particularly Burton Faust, Vice-President for Administration, Benton Hickock, Custodian of Pictures, Eugenio de Bellard Pietri, and Alexander Wetmore, until recently Secretary of the Smithsonian Institution, who, since the early days of the society, has offered it encouragement and good counsel.

The owners and operators of more than one hundred caves which have been lighted and improved so that the public may safely and comfortably enjoy their superlative features have extended many special courtesies over a period of two decades.

The United States National Park Service, the custodian of eleven caves which are open to the public, has helped in obtaining photographs and data. Particular mention should be made of Bennett T. Gale of the National Park Service, R. Taylor Hoskins of Carlsbad Caverns National Park, and Perry T. Brown of Mammoth Cave National Park.

Finally, the editors express their warmest thanks to Peggy Mohr and to Lucille Sloane and Bruce Sloane, for their many suggestions and for hours of typing and correcting manuscripts and of proofreading. The continued personal interest, advice, encouragement, and skilled editing of Helen Stewart are gratefully acknowledged.

September, 1955
New York City

CHARLES E. MOHR
HOWARD N. SLOANE

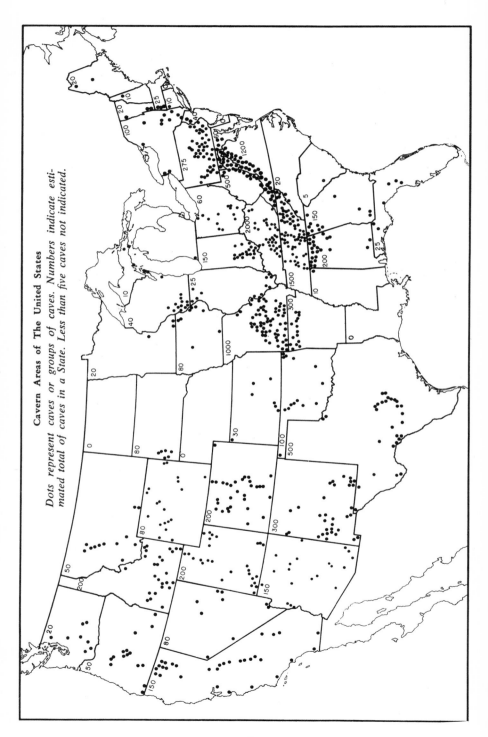

Cavern Areas of The United States

Dots represent caves or groups of caves. Numbers indicate estimated total of caves in a State. Less than five caves not indicated.

x

Contents

Foreword *Alexander Wetmore*	v
Editors' Note	vii
Introducing the American Underworld *William E. Davies*	3
The Devil's Sinkhole *Patrick J. White*	19
Cave of the Vampires *Charles E. Mohr*	39
Assault on Schoolhouse *Ida V. Sawtelle*	59
The Impossible Pit *Roger W. Brucker*	78
The Miners' Bathtub *William R. Halliday*	90
Mammoth Cave's Underground Wilderness *Henry W. Lix*	105
Medicine, Miners and Mummies *Howard N. Sloane*	116
The Valley of Virginia *William E. Davies*	130
Carlsbad Caverns *T. Homer Black*	142
The Death of Floyd Collins *Roger W. Brucker*	158
Bones and a Railway *Brother G. Nicholas, F.S.C.*	172

Earliest Americans   185
Frank C. Hibben

Hidden Skeletons of the Mother Lode   193
William R. Halliday

Bats and Bombs   206
Charles E. Mohr

... and Bands   223
Charles E. Mohr

The One Who Cries   234
Howard N. Sloane

The Idol of the Cave   243
M. R. Harrington

No Eyes in the Darkness   256
William Bridges

Ozark Cave Life   269
Charles E. Mohr

The Leather Man   290
Leroy W. Foote

The Cave in Rock Murderers   304
George F. Jackson

Mark Twain Cave   313
Howard N. Sloane

Lost Footprints   317
George F. Jackson

Index   325

CELEBRATED AMERICAN CAVES

Introducing the American Underworld

An assistant branch chief of the United States Geological Survey, William E. Davies has been a member of survey parties to Little America, in the Antarctic, and to Thule, Greenland. During World War II he was a captain assigned to the Army Map Service.

Davies is president of the National Speleological Society, and at the age of thirty-eight, is recognized as one of America's foremost experts on caves and other underground sites. His voluminous compilation, Bibliography of North American Speleology, 1707-1950, *is nearing publication.*

No armchair speleologist, Davies has explored some twelve hundred caves in the United States, and a few of New Zealand's enormous and spectacular caves. He has organized a number of tours of caves for non-cavers, and is able to share the exuberance of the neophyte spelunker while satisfying the disciplined curiosity of the scientist.

WILLIAM E. DAVIES

One warm summer night in 1948 we were relaxing on the front porch of the hotel in Franklin, West Virginia. Supper was over, and we were enjoying our evening in the usual manner—gossip, stories, or just plain talk. A building supply salesman, two or three tourists, a county judge, and several others sat there, rocking to and fro. The conversation got around to caves, as it seemed to quite often because Tom Richards and I were engaged in a study of West Virginia caves for the State Geological Survey.

We were well prepared for the first question—one we knew was sure to come—and it was the salesman, a practical man, who asked, "Why do you fellows explore these caves; any money in it?"

It wasn't a hard question to answer, for often the best reply is the one Mallory made when he was asked why he wanted to climb Everest: "Because it is there." Caves too are a challenge simply because they are there. But there is really more to it than that, for there is the possibility of finding deposits of economic value—not metals, but sands and clays that industry needs. Or, we could have enumerated a list as long as your arm about the value of caves to man: water supply, engineering properties of rocks, molds and other fungi of possible use as antibiotics, bomb shelters, and many others.

One question always leads to another, and the judge got his in next, "Where do caves occur?"

I picked this one as a chance to review the subject. "First there are the rocks. Caves are found in several kinds of rocks. One group of caves, the ones we call solution caves, are developed in rocks that, like limestone, are relatively soluble in water. They are fairly widespread in much of North America. The limestones are formed by calcium carbonate and were deposited by chemical or organic precipitation in ancient seas. They are sediments, and as such they are bedded or layered. In some cases the layers are paper-thin, while in others they are many feet thick.

"Another of the soluble rocks is gypsum," I went on. "It is formed of calcium sulphate and, like limestone, is deposited in beds or layers. It is precipitated by chemical activity that accompanies the evaporation of dying seas. Large caves sometimes develop in gypsum, but they are not so common as limestone ones because gypsum is less widespread. There are other soluble rocks, but for practical purposes they can be passed over as far as caves are concerned."

The conversation continued, and we discussed other kinds of caves. Most common, after the solution caves, are the ones found in volcanic areas. Developed in lava, they are tube-like in form. Some of them are tremendous, ranging up to several miles long and thirty to fifty feet high and wide. They are unbelievably numerous in regions like northern California, where, in Lava Beds National Monument, 297 lava caves have been explored.

We spoke, too, of sea caves. Wave action along unprotected cliffs often erodes the weaker zones in the rock, producing deep fissures and grottoes like those along the coast of southern California. We went on to caves formed by the openings within piles of huge boulders, like the Polar Caves in New Hampshire, and the rock houses or shelters worn in soft, weak rocks, sandstones in particular, by the erosive forces of wind and water—caves like the famous Indian cliff dwellings in the Canyon de Chelly, in Arizona, or many others in the Southwest.

In many caves nature has provided her own perpetual ice box. This phenomenon is caused by the gentle settling into the cave of cold, subfreezing winter air that freezes the water already present and any that may enter later. The chilled rocks provide an insulation which preserves the ice for much or all of the year. These ice boxes are developed most commonly in lava caves, sometimes as far south as New Mexico and Arizona. The finest examples are to be found in the Lava Beds National Monument, California, in Oregon's Lava River State Park, and notably in and near the Craters of the Moon National Monument in Idaho. Permanent ice may also be found in solution caves in limestone and in the talus caves formed by hollows beneath huge piles of boulders, or in wells and mines.

During our conversation it had grown dark. As in all small towns, the passers-by casually joined in the conversation. The local dentist spoke up. He had often ventured a short distance into a few of the caves near Franklin, he said, and was intrigued by what he had seen. Wisely, he had not gone alone much beyond the entrance and was waiting for a chance to explore farther with someone who was interested. We talked at length about adventures in caves—stories of men lost in caves, and great feats of cavern exploration . . . and of prehistoric man and his cave art. Inevitably the conversation brought on the all-important question of how caves come to be.

There are currently two main schools of thought on the subject, I explained. One group believes that the cavern passages are developed within a zone which lies between the surface of the earth and the water table, the depth at which the ground is saturated with water. In this zone water moves vertically downward from the surface and also sideways along the top of the water level.

This moving ground water dissolves relatively large quantities of limestone as it traverses the rocks. Small tubes are formed and in time are enlarged into passageways and eventually into rooms. In this way the ground water acts somewhat like a surface stream as it develops and enlarges its course. At times when air occupies the solution channels, water seeping downward from the surface deposits the beautiful formations that often decorate cave passageways.

The judge detected a flaw in the theory when he asked how it could account for a cave he had been in recently that was developed

Solution pendants or channels in Laurel Creek Cave, West Virginia, are examined by speleologist Nancy Rogers. These are called anastomoses. *William E. Davies*

as a maze, with numerous uniformly spaced, intersecting passages, equally developed in two directions. Such caves are like city streets in plan and obviously could not be developed by an underground stream since such streams would select certain openings to the exclusion of others.

I could only acknowledge the failure of the theory to account for this, and proceeded to explain the second of the schools of thought on cavern development. According to this concept, there are two distinct stages: The first stage takes place at random depths within the zone where the ground is permanently saturated with water. Here the water circulates slowly, and since it fills all the openings, it is possible to develop passages in any direction, singly or in groups. During the long geological periods involved, extensive caverns can be developed.

With the gradual wearing away of the surface of the land as valleys become deeper and deeper, the cavern passages are drained.

Fractured rock faces, characteristic of a fault, seen in Wind Cave, Pennsylvania. *Charles E. Mohr*

They now lie above the saturated zone. In this second stage the caverns are filled with air, and mineral-bearing water seeping through the rocks deposits stalactites and other beautiful formations in the cave.

I hastened to add that a minor interval is recognized at the close of the first stage. At this time, when the development of openings within the zone of saturation is drawing to a close, layers of clay, sand, and at times gravel, may accumulate in the passages, in some cases completely filling them.

The group felt that the second theory accounted for conditions fairly well, and we discussed it at some length. The dentist, however, asked just how it applied to caves around Franklin. As with the other theory, there was a snag. The rocks here, as in the rest of the Appalachians, are steeply tilted and crumpled. According to the theory of the circulation of water in the zone of saturation, one might expect caves to be formed in the more soluble beds of rocks. But this is seldom the case in the Appalachians, where most caves have relatively level passages that cut across the tilted bed of rocks with little regard to the varying solubility of different strata.

The judge sagely observed that probably the truth would turn out to be a compromise incorporating certain points of each theory. I agreed, pointing out that recent investigations were bringing us to this conclusion.

It was very late, and we had a big day of caving ahead of us. Before heading for our room we extended an invitation to the group to join us in the morning. The dentist jumped at the opportunity and asked if he might bring his two teen-aged children. We'd be delighted to have them, we said.

The next morning we met the dentist and his boy and girl at the cave, and after lighting the carbide lamps, we began our subterranean tour.

The cave we had chosen could not be called typical, any more than any other, for each cave possesses certain distinguishing features. Our cave contained several different types of passages. For the most part they were narrow openings high enough to permit walking. The floor had its ups and downs and was covered in places by fragments of rocks; in other places there was sand or clay. Here and there along the way, the passages expanded into large rooms measuring two or three hundred feet in length, about half that wide, and with high domed ceilings fifty to seventy feet above the floor. In a few spots the passages were so low that we had to crawl, and at two points there was a tight squeeze that elicited considerable grunts and groans.

Deep, well-like pits that are common in many caves were lacking in ours, but dome pits, large circular openings in the form of inverted wells, extended upward into the rock above certain passages, some attaining a height of almost a hundred feet. Water dripped continuously from these dome pits and splashed in small pools on the floor of the passage.

Near the rear of the cave were several side passages in which formations were numerous. From the ceiling hung clusters of white stalactites, some massive and bulbous, others thin and hollow like soda straws. These were the result of seepage and drip of water from the ceiling. The water, slightly acid from its contact with air and

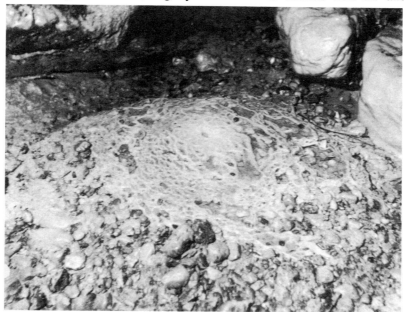

The beginning of what may become a large column of dripstone, Laurel Creek Cave, West Virginia. *William E. Davies*

soil, had dissolved calcium carbonate from the limestone above the cave. As the solution oozed through the ceiling it hung in drops, poised for a time until their increasing weight caused them to fall to the floor. Before they fell, evaporation or the loss of carbon dioxide from the solution caused some of the mineral to be left on the ceiling. In time the minute deposit elongated to form a soda straw, then enlarged into a more typical stalactite.

We noted that the stalagmites on the floor of the cave were paired with stalactites above and were obviously formed where the dripping water carried material in solution to the floor; here it precipitated as evaporation occurred. We moved along the passage, stopping to examine a sheet of crystalline calcite formed where a thin film of water flowed down the wall.

Falling drops leaving minute deposits of calcium carbonate eventually may build massive stalagmites such as this one in Cave-Without-A-Name, Texas. *Zintgraff*

Mixed in with the stalactites were a number of erratic formations. Some segments of the stalactites had developed in apparent defiance of gravity, for they grew horizontally or turned abruptly upward. These twisted formations were helictites, and their structure is attributed to changes in orientation of certain physical properties of the crystal as it grows.

As we were retracing our way, the dentist remarked that it was easy to see why people visited caves. "Formations are not so prolific in most caves," I said. "Often there are formations here or there, but in general, bare rock is encountered most of the time."

When we arrived back at the main passage, all the members of the party had the same question to ask—was there any more to see? They were assured there was more . . . much more. As a matter of fact, there was a whole new cave lying at another level, fifty feet below the one we had traversed. We reached this lower level by way of a steep clay slope that ended in a sheer drop for the last ten feet. Even so, this connection between levels was much easier to

Helictites in Lower Salt River Cave, Tennessee. These grotesque cave formations grow in defiance of gravity, assuming weird, twisted shapes. *Charles E. Mohr*

Unusual, white helictites in Ohio Caverns, Ohio. *Charles E. Mohr*

Beautiful pure white calcite stalactites and stalagmites in Ohio Caverns, Ohio. *Charles E. Mohr*

negotiate than the pits or wells that commonly connect different levels.

The lower level was like a different cave. Gone were the dry walkways we had traveled. Down here water was almost everywhere. A stream several inches deep and a couple of feet wide flowed along the floor. Here and there it formed a pool that just about covered the floor. At one point it plunged over a falls, the roar of which in the confined passages seemed to rival Niagara. The stream was intriguing, but it didn't compare with the celebrated Echo River in Mammoth Cave, Kentucky, on which boats travel, nor was there a large lake in a vaulted gallery as has been discovered in some caves. We followed the stream until we were stopped abruptly by the ceiling dipping sharply beneath the surface of a small pool.

We lacked the diving equipment we should need to penetrate beyond this point, and we turned back, taking a quick look into the side passages of this lower level. Finally we made our way back to the entrance. Sitting at the mouth of the cave with the late afternoon sun casting its rays deep into the recesses of the cave, we resumed our leisurely discussion of cave features.

There were several questions about moisture and temperature. "In general," I said, "the temperature well inside a cave is the same as the yearly average for the area would be. In the Central and Middle Atlantic States this ranges from 40 to 60 degrees Fahrenheit. In any one cave it varies little, if at all. Humidity in caves, except in those with considerable quantities of gypsum, generally is close to 100 per cent.

A steady circulation of air is evident in most cave passages. Sometimes this movement is so strong that air blows out the entrance with great force, giving rise to the designation "blowing cave." In a few caves the direction of the current of air reverses for short periods so that the air is alternately blown out and sucked in, forming a "breathing cave."

We were well into this discussion when suddenly we became aware of a cool breath of air flowing over us from the mouth of the cave. Nature's air conditioner, the world's first, cheapest, and most durable, was making an otherwise hot, muggy afternoon pleasant and comfortable.

We had missed seeing any mineral deposits in this cave. Had we looked a little more closely, however, we might have noticed one or more of a number of well-developed mineral forms which are present in most caves: manganese dioxide in the form of black, soot-like coatings on the walls or as black stains or encrustations on pebbles, rocks, and clay layers, gypsum in crystalline form in cave fills or as formations, calcium nitrate in the niter deposits, and the most common of all, calcite, which makes up most of the formations. In certain cave regions we might encounter the lead mineral, galena, and zinc in the form of sphalerite; fluorite also is common.

The cave earths or fills in which minerals commonly occur consist of relatively thick layers of clay, sand, and gravel. They are present in most caves, and their former depth is indicated by the presence of gravel or other fill material in niches or grooves high above the existing floor. Fills are often the burial ground of ancient animals, and many important paleontological finds have been made in them.

In many caves the explorer encounters great blocks or slabs of rock which piled up when portions of the ceiling collapsed. In some places this breakdown is local, extending but a hundred feet or so along a passage. However, this type of breakdown often extends upward for fifty or a hundred feet into the ceiling and the enormous heap of broken rock and rubble inspires caution and respect even among the most experienced explorers. In another form the break-

Gravel and clay fill in Schoolhouse Cave, West Virginia. It is in such fill that artifacts and fossils are sometimes found. *William E. Davies*

down extends for great distances along a passage but involves only a few beds of the overlying rock.

Rockfalls in caves are similar to collapses in mines, but none are known to have taken place since systematic studies of caves commenced almost a century ago. Apparently cavern passages are in a reasonably stable state.

It was late now, and our thoughts were turning to warm baths and hot meals. Our trip had given us an idea of what a cave is like. Pleasantly weary but exhilarated by glimpses of a strange and beautiful world, we walked slowly back to our cars and headed for town. Some of the group would, I knew, want to go again—go farther—into this fantastic and exciting territory. There were a few notes of caution to pass along.

Whether cave exploration is safe or dangerous depends largely on the individual caver. Under no circumstances should one enter a cave alone. Always have a partner or preferably two or three companions. In all exploration stay well within the bounds of your en-

Fallen rock, called breakdown, in Laurel Creek Cave, West Virginia. *William E. Davies*

durance. Limit your climbing efforts to your proved climbing abilities. Let someone outside the cave know where you are.

Be respectful of the cave owner's rights. Request his permission, in advance if possible, to enter his cave. Honor the conditions he stipulates.

Last, and yet of great importance, is the matter of conservation. Be careful to leave the cave and all its contents as you found it. Leave formations and animal life untouched. Stay on the existing pathways, and do nothing to mar the walls and ceilings. Remember that it took thousands of years to create the wonders you enjoy. Don't permit a few moments of thoughtlessness or carelessness to rob future generations of the same pleasure.

Our cave was only one of the five hundred known in West Virginia, one of the estimated 30,000 in the United States, and one of probably 500,000 in the world. The 30,000 caves in the United States are spread over a number of areas, and each state, with the exception of Delaware, is known to have caves. In the eastern United States, there are extensive cave areas long known to travel-conscious Americans. The broad limestone lowland on the eastern side of the Appalachians extending from New Jersey to Alabama has numerous well known caves, mostly of moderate size, seldom extending for more than a couple of thousand feet. The caves are in intricately folded limestones and the pattern of passages is irregular, yet simple. Many of them are beautifully decorated.

To the west of the Appalachian Valley is an area of long, parallel ridges and valleys, the Appalachian Mountains. Here are many hundreds of caves, and frequently they are large, some up to a mile or more in length. Patterns of these caves are geometrical with the passages following systems of joints in the rocks.

Typical cave country surface terrain in West Virginia, showing the limestone outcroppings called karst. *William E. Davies*

Introducing the American Underworld

To the west of the Appalachians, in the plateaus of central Kentucky, Tennessee, Alabama, southern Indiana, and parts of southern West Virginia lies America's real cave country. In this area the caves are measured in miles instead of feet, and the interconnected complex of passages has perforated the earth until it resembles a block of Swiss cheese. The caves are in very thick, flat-lying limestones that are extremely soluble, and the passages often are comfortably large and broad with openings forty to fifty feet high and wide, extending for a mile or more. Gigantic pits and domes over a hundred feet deep or high intersect the passages and are more common in these caves than in those of other areas.

In the cavern area of north and central Florida mammoth underground streams drain the area, and some boil up from the ocean floor miles from land. Dry caves that can be entered are mainly in the northern part of the state. These caves are small or moderate in size.

The caves in the Ozarks, in Missouri and Arkansas, have long been noted in folklore. They are very numerous, of moderate size, and often are associated with large springs and underground streams.

In the West the caves are less numerous but of great variety. In the Rockies, the Big Horns, the Black Hills, and related mountains, the folded, cavernous limestones lie along the flanks of many ridges. Caves are moderate in size and, like those in the Appalachians, are generally geometrical in pattern.

In the Pacific Northwest, the Snake River area of Idaho, and the Bend Country of southern Oregon and northern California, are vast lava fields with their numerous tubes or caves. The rough lava terrain is similar to that of cavernous limestone or karst and because of this resemblance is often referred to as pseudokarst.

Cave country in western Oklahoma and adjacent Texas is gypsum country. Here thick beds of gypsum have been subjected to solution and numerous small or moderate-sized caves exist. To the south and west, the Edwards Plateau of central Texas is a large area of thick, flat-lying cavernous limestones. Beneath this surface are numerous caves. Some are large caverns, some have deep, vertical sink entrances. Similar rocks with vast caverns are found in southeastern New Mexico, where one of the world's largest caves, Carlsbad Caverns, is found.

The deepest caves yet discovered have been encountered in oil wells in west Texas. Here the driller's bit has tapped vast reservoirs of oil in caves that rival Carlsbad in size. These caves are over 9,000 feet below the surface and are fossil caves, for they were developed in relation to a land surface that existed over 100,000,000 years ago. After the caves were formed they were buried under thousands of feet of sediment and gradually filled with oil, forming some of the largest oil fields of the United States.

American cavern areas extend into the tropics in Mexico and Guatemala, in the West Indies islands of Cuba, Puerto Rico, Dominica, and Jamaica. Here solution has worked freely on the limestone to such an extent that it has been said that the land is a lot of caves held together by a little rock. These caves are large and often beautifully decorated. Some in the mountainous areas are very deep.

Canada and much of the north country was dealt a telling blow as far as caves go when the glaciers of the Ice Age swept over the land. They gouged away the caverns or covered and filled them with debris. Only in the gypsum beds of Nova Scotia, New Brunswick, and British Columbia, and in the limestones of the St. Lawrence Valley and Ontario Lowlands, are caves known to exist. They are generally quite small and simple in plan.

Caves often are encountered deep within the earth in mining and drilling operations. Generally these caves are large openings without apparent channels connecting them to other cavities. The caves may be lined with numerous, gigantic crystals of calcite, selenite (gypsum), and often with ore minerals of copper, zinc, and lead. Such mineral caves have been encountered in mines at Naica, Mexico; Bisbee, Arizona; the lead and zinc districts of Missouri; and the Mississippi Valley area of Wisconsin and Illinois.

The Devil's Sinkhole

The past twenty years have seen a great increase in cave exploration. The use of new mechanical aids, lightweight ladders, concentrated foods, more powerful lights, collapsible boats, and diving equipment have made possible successful invasions of caves heretofore impregnable or forbiddingly hazardous. As important as the development of new equipment and reliable means of communication underground, are the increased, widespread skill in the art of rock-climbing, and the effective planning and teamwork which precede a difficult cave investigation and govern the conduct of the explorers.

Pat White's first descent into the Devil's Sinkhole was made in the "early days" of organized caving, before modern safety procedures had been widely adopted. He probably would chafe under the limitations placed today on the members of exploring teams, should they tackle a cave of the dimensions and hazards of this vast sinkhole.

His whirlwind, state-wide explorations, often with only a single companion, did much to focus attention on the speleological wonders and wealth of Texas. Of all those wonders the Devil's Sinkhole is perhaps the greatest.

Patrick J. White

We saw it suddenly towering against the starry sky, an undulating plume of blackness seventy-five feet thick and rearing one hundred and fifty feet above the prairie. It seemed to pulse from within, as though with a life of its own—a massive, wavering column, inexplicably held fast to a stationary base.

Intuitively, without consciously identifying it, we linked it to the Devil's Sinkhole. But not until its overpowering odor reached us did we realize that the whole plume was composed of flying bats. For some reason, the bats did not fly off as soon as they rose above ground

Ralph Velasco being lowered to join Floyd Potter, two hundred feet below in Devil's Sinkhole. Potter is standing on top of a huge mountain of rock. Ed Raney lies on the ledge overhanging the hole, while Pat White operates the auto to which the cable is attached. *Tom Culverwell*

level, but continued their rotary flight within the column. Perhaps it was because air directly above the Hole was warmer than that surrounding it.

Without their warning, we might have stumbled into the huge mouth of the pit, which is unmarked and perfectly flush with the ground. As we approached, the rocky soil underfoot gave way abruptly to great slabs of crumbling limestone. And we found ourselves staring into the vortex of that mass of swirling blackness, deep down into the earth. We lay flat on our bellies and with great care crawled until our heads were out over the shaft. The beating of wings was all about us, punctuated by the shrill shriek of bats.

Our lights picked up the opposite ledge of the opening seventy-five feet across from where we lay. And, as we lowered the beams, we saw that the ledges all around were deeply undercut and that we were upon an overhang to a shaft far larger than its opening.

About seventy feet down, another ledge sloped precipitously outward, only to dive inward and out of sight as it was undercut. Forty feet lower still, the ledging repeated itself. And another forty feet down, our lights picked up three boulders directly in the center of the Hole.

We thought then that these boulders were on the bottom. We still thought so the following day, when we viewed them from the top in the full light of the sun. We were wrong. But it is only from inside the Hole that one can see that they rest upon the peak of a huge, conical mountain of tumbledown, which pitches downward another two hundred feet to a three-hundred-foot-broad base out of sight beneath the deepest overhang.

For a long time we lay on the edge of the pit watching the bats, which, oddly enough, made no move to leave their swirling column. Then, alternately silent with awe and talkative with enthusiasm, we trudged back across the prairie to our automobile. We were tired. We had been eight hours late in reaching the Hole, because of equipment difficulties which had beset us since early morning.

With no way of foreseeing the dimensions of the pit or the unique problems a descent into it would involve, we had equipped ourselves with 1,000 feet of three-eighths-inch cable, cable clamps, a snatch-block pulley, and a bosun's chair, rigged by Ralph Velasco, who, because of several years' experience as a mate in the merchant marine, had been put in charge of the tackle.

There were also, of course, the usual one hundred feet of safety rope, assorted lights, collecting equipment, and other gear. Velasco advised a more complete array of lowering equipment, but capitulated to an argument based upon expenses. Before we were done, all of us wished fervently that we had taken his advice. Each of us went through some nightmarish moments, and we wasted so much time on mechanical details that it was only after a second trip, the following weekend, that we were satisfied with our exploration.

We had planned to leave San Antonio at nine that morning, and reach Rocksprings, about 150 miles away, by noon. Last-minute difficulties in securing our equipment delayed our start until one o'clock. And a broken fan belt on Floyd Potter's car marooned us twenty miles from Rocksprings on the desolate prairie. Two hours were lost while Eddie Raney flagged a ride into town and returned with a new belt. We spent this time searching for herpetological specimens for Floyd, but rock after rock, upturned at considerable effort, yielded only nests of ants and dens of writhing scorpions.

We arrived in Rocksprings about an hour before sundown and were sent back ten miles along the way we had come to the Diamond Bar Ranch, where the townspeople said we could get directions to the Hole. At the ranch house, a Mexican foreman greeted us alertly and in passable English. But when we asked how to get to the Sinkhole, his manner abruptly became hostile, and he lapsed into that nearly unintelligible patois colloquially called "Tex-Mex."

The roads were impassable, he said, and the Hole was so far out on the open range that it could not be found without a guide. He added hastily that there was no one available who could serve in that capacity. Nothing we could say moved him until Ralph Velasco addressed him in Spanish.

Velasco's idiom is authentic, and the foreman warmed to him at once. In an excited gush of words, the foreman told him that the Sinkhole was evil—that it was a place of ghosts and devils and terrible legends—that it was the Mother Nest of all rattlesnakes. And he swore that nothing in the world could force him to go near it.

Ralph, with a native understanding of these people, did not laugh, but sank to one knee and while scratching the ground with a stick, began to talk seriously, softly, with the foreman. The sun sank out of sight beyond the horizon before we had our information.

Turning south from the highway through the first gate west of the Diamond Bar, we immediately plunged into a bog, but fortunately the car was equipped with mud tires. After we had slithered through the bog for about a mile, it gave way to a hard, rocky trail, as bad as any I have ever seen. The car took a terrible beating. After about two miles of this rockway, we reached a fork in the road which had not been mentioned in our instructions. There we split the party and continued on foot. Eddie Raney and Floyd Potter took the northern branch, Ralph Velasco and I the southern. Even the walking was difficult that night, so rocky was the ground.

Gradually the trail disappeared, and shortly thereafter we saw Eddie's light glimmering in the mesquite to our left. We called and they joined us, having circled up parallel to our course.

We debated the wisdom of abandoning the search until the next day, but the urge to see the fantastic abyss was too strong. Furthermore, the night was relatively warm, and the rattlesnakes were in hibernation. Never have I seen brighter stars.

We moved onward carefully. And then, we saw the monster plume of bats swirling above the Hole!

I shall never forget that first sight of the Devil's Sinkhole, or one moment of that night of brilliant stars, whirling, stinking bats, and that yawning gulf at our feet. There was a weird majesty in the moment, and not one of the four of us escaped the emotional impact of it. It lingered into our dreams and caused us a dozen subconscious delays in the morning. It was seven-thirty before we had finished breakfast at the Rocksprings hotel and nine o'clock before we were back at the Hole. We had planned to be there at dawn.

In the full light of day, the Hole lost much of its eeriness, but none of its impressiveness. And as the sun mounted in the sky, the old, familiar passion for exploration replaced the awe we had felt the night before.

The daylight reached down into the huge crucible all the way to the cone of the interior mountain, and by some strange trick of perspective it made this point, which towers some two hundred feet above the cavern floor, appear to be the floor itself—perfectly level and only about 150 feet below us.

Stalactites so massive as to resemble parts of the ledges themselves dropped downward, covered with moss. Water dripped in a fringe from an upper tier, lost itself against the wall after about thirty feet, and could be heard gurgling somewhere near the bottom of the shaft. Gray-green moss and lichens grew thickly on the weathered limestone. Some eighty feet from the surface a mass of ivy hung like a giant, dark-green tassel against the wall. On the precipitous slope of the middle ledge, a tree, long dead, still thrust its bare branches upward.

The shaft itself, when viewed from above, did not reveal its true dimensions. It seemed to fall sheer from ledge to ledge. Actually, it tapers on all sides continually outward, away from its axis.

From a small hole, artificially cut in the surface ledge, an ancient ladder, now rotten and crumbling, reached downward to the first ledge. It was held in place by a rust-covered cable, and its extension to the second ledge had entirely fallen away. Another small, artificially carved hole near the latter entrance once served as a port through which guano was hauled to the surface some twenty years ago. The mining venture soon ended in failure.

A half-century ago an enterprising rancher dropped his windmill pipes to the Emerald Lake in one of the caves at the bottom. Nothing is known of the method by which he gained entrance and egress, or the manner in which he performed the heroic feat of laying his pipes to the surface.

Certainly, as we looked into the Hole that morning, no one of us would have considered such a task without professional aid. Yet, we even then underestimated the difficulty, not so much of entering, but of returning from its depths. We set to work assembling our tackle with a lightheartedness not justified by the materials we were using.

We attached the bosun's chair (a device that looks like a swing suspended from one rather than two lines) to the end of our three-eighths-inch cable, weighted the chair, and paid out the cable until the bottom of the Hole had been reached. We then backed the two-door sedan as close to the edge as safety permitted and affixed the free end of the cable to the bumper. The clamps were tight and the chair solid. We had no fear of the line.

The pulley had been abandoned in favor of letting the cable run directly against the edge of the limestone ledge, which was obviously soft enough to permit a track to be cut by the action of the cable. We made two trial runs with the car, marking the path with rocks and building cairns at the points where the car should be slowed for passage of the ledges, and where it should be stopped when the bosun's chair had reached the bottom. It all looked reasonably satisfactory even for swinging out that tremendous distance over the jagged rocks below.

We wanted Ralph to remain topside and watch the rigging on the first descent. It was Eddie's first experience with any shaft deeper than a hundred feet. So the maiden drop was left for Floyd or me. We flipped a coin, and I lost. While Floyd crawled into the bosun's chair, I took the car to the end of the run. Eddie stationed himself midway between the car and the ledge to relay signals, and Ralph remained at the ledge to watch the line and to give directions to Floyd as he began backing out over the brink. When the bosun's chair swung free, Ralph signaled, "Lower away!"

At exactly 10:55 A.M., I got the signal. At 11:00 Floyd touched the first ledge, and, while dangling in the swing against its steep face, began trying to hack down the dead tree that was in the path of our descent. At 11:01 he gave up the task. At 11:03 the wheels of the car reached the cairn marked "on the bottom," and Floyd's voice drifted up the shaft. "Everything's okay. Pull up the chair!"

I leapt out of the car and looked over the lip of the shaft. With Floyd standing ant-like among the boulders at the bottom, I began to perceive for the first time the true depths of the Hole. After a moment Ralph turned away from the pit. "Well, let's go," he said. It had been decided that he was to join Floyd.

I got back into the car and drove forward to the cairn marked "on top." Eddie helped Ralph over the ledge, and I began lowering at 11:20. Ralph was on the bottom at exactly 11:25.

Once again I went to the edge of the Hole and watched the two men below. "We're on top of a hill," one of them shouted up. "This thing is much bigger than we thought. Drop a line. We'll measure it."

Eddie threw them a line, and they disappeared under an overhang with the end of it. The illusion was that they had traveled over a perfectly level surface until the overhang cut them from view; actually their path had been down a 45-degree slope.

After a short while Ralph appeared and called us to haul in the line and measure it. Later we would triangulate from this measurement and from others taken in the Hole. This calculation indicated a depth of 407 feet sheer from the surface of the water level—a figure which we believe to be very nearly exact.

We sent down a load of equipment, and, leaving the chair on the floor of the cave, Eddie and I sat down to wait while our friends explored. They had promised not to be gone longer than four hours.

It was past noon, and the sun was so hot that we soon spread a blanket on the flat, limestone ledge at the pit's entrance, stripped off our clothes, and stretched out for a sun bath with our heads not more than a foot from the top of the shaft. We fell asleep and rested comfortably until three o'clock, when we heard shouts from below. I don't know what the exact temperature was, but during the bat flight the preceding night, I had noted it was 38 degrees inside the car. I mention this because of the probability that the weather conditions dictated the peculiar activity of the bats that weekend.

There in the sun, it did not seem that it could have been so cold the night before. But the temperature dropped rapidly at twilight, and the bats came out again that night. Exactly at sundown they began their spiraling in the huge shaft. Slowly, as more and more bats joined the flight, the whirling column began to extend upward into the air above the Hole. By eight o'clock, it seemed about the same height as on the preceding evening. But this was the last night we saw the phenomenal column.

During the next week, a norther howled down out of the Panhandle. And when we returned to the Hole, the following Saturday, there was not a sign of a flying bat anywhere. Nor did they emerge the next day, Sunday. Apparently, the marked difference in the temperatures of the two weekends accounted for the change in their activity, for they did not come out. On both days of the second weekend, we saw them hanging torpidly in their usual spots under the ledges.

M. E. Stone, a member of the Rocksprings Chamber of Commerce, who had observed the bat flight in summer, told us that their customary procedure was an upward spiraling, followed by a mass break for the southeast. He said that practically all the bats flew off in this direction and that there were two flights every night. One was at about 5:45 P.M., and the other approximately one hour later. He added that each lasted about forty-five minutes and was preceded by a few bats which, in his opinion, appeared to be "scouting," because they returned down the Hole before the flights began. On the one weekend during which we observed bat activity, there was nothing to indicate the direction the bats would have taken had they broken from their column. Nor did we see two flights.

Floyd Potter secured two distinct species of bats in the Hole! One was the usual Little Brown Bat, the other was the Long-eared or Lump-nosed Bat. In all probability, however, neither of these bats was of the species that made up the flight we saw. It seems certain that the millions of bats here are the Free-tailed variety, *Tadarida mexicana*. Since the Hole is seldom visited by the natives, we were unable to find anyone who had made other than casual and very occasional observations there.

Potter succeeded in taking from the water at the bottom of the cave a number of isopods. We saw no other cave fauna, although Potter netted zealously through a good many pools. He found a hibernating lizard under a stone about a thousand yards from the Hole. Aside from these, the only fauna observed by us in the vicinity were a number of flies, presumably the common Bluebottle Fly.

A descent on the improved rigging is supervised by Captain F. M. Johnson on a later trip to the Sinkhole. The car in the foreground serves as a mooring for the trolley. *Francis Wood*

As soon as we heard Velasco and Potter calling up the shaft, we dressed and manned the rigging. Floyd climbed into the bosun's chair, and I brought him up to the highest of the two interior ledges without difficulty. There he tested the lower rungs of the old ladder which led down from the manhole at the top. Finding the ladder in very dangerous condition, he made another unsuccessful attempt to chop the tree out of our path. And then he signaled to be hauled out.

There the trouble began. I had proceeded with the car only fifty or sixty feet when I began to notice an irregularity in the pull of the cable, which manifested itself as an almost imperceptible jerking. Just as Eddie called, I stopped the car and ran back to the pit. Floyd was twirling gently on the end of the cable about a hundred feet above the boulder-strewn floor and fifty feet below us.

At the edge of the pit, I discovered that friction had worn a groove in the soft limestone where the cable paid over the rim. On the downward trips, this groove had served as a satisfactory guide. But by now it was slotted so deeply that the cable could no longer run smoothly. It was feeding in slight jerks, and to make matters worse, was so deeply wedged that we were unable to lift it out of the slot and give it a new seat. Our attempts only made Floyd twirl more giddily on his airy perch.

Ralph had warned us that a cable, which will pull apart only under great stress, will snap with relative ease. So we were very apprehensive about the jerking, which we knew would get progressively worse as the slot wore deeper. We decided to drop Floyd onto the ledge and clear the bearing. Very slowly, I began to back the car. Eddie's shout stopped me. The cable behind the car had gone slack. Floyd was still dangling in the same place; his weight was insufficient to pull the cable through the slot. There was nothing to do but go forward.

With great care, we inched the car ahead and took up the slack. Then, as a cloud of white powder rose from the slot, the chair began to rise. I was driving so slowly that it was necessary to slip the clutch constantly. Suddenly the car began to skid—at least that is what I thought was happening.

I released the clutch fully. The motor apparently engaged. But the car did not move. Not until I smelled burning rubber did I realize that it was the clutch which was failing to hold and not the tires.

In a wave of burning rubber we would proceed a few feet. Stop. Race the engine. Lose a little ground. And then inch forward again. I recalled the rough treatment the car had taken going through the bog and over the rockway from the Diamond Bar Ranch. After a seemingly interminable period, the car's front wheels reached a spot only a foot from the cairn which marked the final stopping place. That last foot was the most difficult.

The ledge at the top of the Hole was only about three feet thick, and the chair had to be brought up as far as possible in order to get the man out and safely over the ledge. At the same time, we could not risk taking the car an inch too far, for that would throw the cable clamps against the slot, and if any appreciable pressure from the car were then applied, the clamps would probably be stripped from the cable—or the cable would snap. Somehow we made that last foot. I set the brakes and gears and ran back to the Hole.

Only Floyd's blackened, sweat-smeared face and the tops of his shoulders were visible above the ledge. His expression was as close to being anxious as I have ever seen it, and his temper, as we worked him upward, as close to being short as it probably ever gets. He had good reason. After that nightmarish rise out of the depths, he was stuck at the ledge, unable to get over the final three feet to safety.

Here was another problem we had not foreseen. The thickness of the ledge, the depth of the undercut, and the set of the tackle combined to wedge the knees of anyone sitting in the swing under the ledge. And there was no way of getting a straight upward pull. We could not lean out over the edge and pull him up because there was nothing to give us purchase. And the cable clamps were too dangerously near the slot to try lifting him by running the car forward.

Floyd was trying to clamber up by sheer exertion, but the angle would not permit it. Finally, we handed him a length of half-inch manila safety line, and he threw a bowline around himself; we made the other end fast to a mesquite bush. This did little to relieve the tension: if we were unable to get him up from his position right at the ledge, what possible luck could we have in hauling him in hand over hand the length of the safety rope?

It took ten agonizing minutes to get him across the ledge, and before it was accomplished he had to wriggle free of the lap lacing and stand upright on the bosun's chair.

I was surprised at his resiliency. He shed his tenseness almost immediately when he reached solid ground. I asked him to handle the car on the next haul we made to bring Ralph up. It was his car, and I thought he might understand its peculiarities better than I did.

Rather nervously, we cleared the cable from the slot, inspected it, and sent the chair down for Ralph. During the drop Floyd stopped several times and experimented with forward runs of the car. By the time the swing reached the bottom, he was confident that the car could effect the lift. Ralph climbed into the chair, and the car started.

Floyd was driving at full idling speed in order to keep the clutch holding. Ralph came up comparatively fast and began to twirl dizzily after passing the lowest edge. Then he came up fast again, spinning like a top. In short order we brought him firmly against the ledge, the last foot of the lift being the only troublesome one.

At this point, however, the situation was exactly the same as before. Ralph hung below the ledge just as Floyd had. We gave him the safety rope, and he made it fast around himself. He was quite nervous, as we all were. The twirling had made him so giddy that he had been forced to shut his eyes and lock his arms around the tackle to keep from falling out of the chair.

Even with three of us topside, it took seven minutes to work him over the ledge. When we did, he lay at full length on the rock regaining his wind and then walked straight away from the Hole, waiting several minutes before returning.

Both men were enthusiastic in their reports of the grandeur of the pit's interior. It had to be seen, they informed us, describing emerald lakes and marvelous formations. Eddie and I decided to go down at once on the tackle as it stood. Most of the trouble had come on the ascent, and Eddie proposed that Floyd and Ralph change the rigging while we were in the cave. He suggested a method which sounded as though it would solve our difficulties on the ascent.

I got into the swing, took the cable in my hands, and backed out over the edge, slithering down until the swing hung free. The car started slowly, and I saw the entrance fade above me. The nearest wall was forty feet away, and the first ledge was coming up fast. I could see that I was going to land in the tree Floyd had been unable

The Devil's Sinkhole

The man is standing on the summit of a mountain which continues downward another two hundred feet into lower levels. The ledge near the man is actually over forty feet above him. The sloping effect is due to camera distortion; the walls are really vertical. *Francis Wood*

to dislodge. I called for a slowdown, and just as I touched the bare topmost branches my speed was checked.

I worked through the tree and began to twirl. Fortunately a thick branch fell under my hand just as I went clear of the tree. I used it to stop the twirling. Floyd had speeded up again, and I came down on the second ledge fairly fast but in the proper position—facing inward. One kick, and I arched outward and clear, over the ledge and down. In a moment I was on the bottom and out of the chair. It was on its way aloft again as I got my first breath-taking view of the Hole from within.

I was standing on top of a mountain of talus, strewn with the biggest limestone blocks I have ever seen. Looking upward, along the jug-shaped shaft to the entrance, which looked like a tiny window against the sky, it was clear that the mountain had been formed by mammoth cave-ins in the dim geologic past. At my feet, this giant hill tumbled downward at about a 45-degree angle for nearly two hundred feet, losing itself in the shadows at the base of the walls of the room.

This was my first full view of the tremendous chamber. It undercuts the lowest ledge so deeply that only the immediate center of it can be seen from the surface. I was instantly impressed by the perfectly smooth walls and the perfect cup shape of the room. It must once have contained a great whirlpool which formed the room as much by erosion as by solution. In this three-hundred-foot-broad cup, probably after the water had drained away, the collapse of ledge after ledge of massive limestone above had formed the mountain and opened the entrance at the surface. Lush green moss grew profusely over all the rocks on the peak of the mountain and provided a fairly comfortable seat on which I settled back to watch Eddie come down.

He backed over the edge with fair dispatch, called for a slowdown to scramble through the tree and over the first ledge. It was his first experience of any kind on lowering tackle, and as soon as he cleared the first ledge, he called, "Okay!" Ralph relayed the call and Floyd speeded up the car.

Eddie began to twirl violently and came down with a rush toward the second ledge. I leapt to my feet in alarm. The lowering crew could not see what was happening! Before I had time to shout, Eddie's spin turned his back toward the ledge, and he slammed down on it with a sickening thud. The blow threw the chair outward, clear of the ledge. I saw Eddie's hands fly off the rigging as he arched out over the hole. His body bent far backward. He was still in the seat, but was held there only by the lacing across his lap. It was not designed to hold a person in the swing.

I thought he surely had broken some bones and perhaps was knocked unconscious and would fall. But he immediately drew himself up in a sitting position, and his hands grasped the rigging again. The speed of the drop suddenly checked, and he was lowered gently to the floor.

All of this happened too fast for any calling back and forth between us. I jumped over a boulder toward him and heard him exclaim: "Ohhh, brother!" That was enough. It did not sound like the groan of an injured person. The crash against the ledge had done no worse than bruise him. He was breathless but otherwise ready to continue the exploration.

The bosun's chair disappeared upward on its way to the re-rigging. But Eddie and I were too busy surveying the weird netherworld to give much thought to our trip to the surface. The mountain fell away from us on every side. The degree of precipitousness appeared from where we stood to vary only slightly, with the easterly slope presenting a somewhat less difficult descent. Slowly we made our way downward in this direction over exceptionally rugged terrain.

Soon we were under the overhang, and the green moss covering the rocks changed abruptly to deep, soft, brown guano. It blanketed everything, filling up the cracks between the rocks and even adhering to the sides of upright slabs. We directed our headlights against the ledge above us, but saw no bats. The guano was often eight inches deep and of a consistency to provide reasonably secure walking. It gave off practically no odor.

As we approached the cup-like walls of the room, it became apparent that the mountain did not end at their base, but continued on down still farther. An investigation revealed that around the entire room the walls came down almost against the sides of the mountain and were there undercut horizontally for varying distances. This undercutting, where not plugged with the talus of the tumbledown, creates the caverns at the bottom of the Hole. In them are the emerald pools of water, the strange crustaceans, and the fabulous formations.

The most exciting view which the pit provides is from a point just outside this final undercut. From there, one looks up nearly two hundred feet along the rocky slope of the mountain, and on upward another 150 feet along the narrowing shaft, and finally outward to the sky. The effect is of looking upward to an open dome in some strange story-book cathedral. The entrance seemed scarcely larger than a postage stamp, and the hanging bosun's chair could not be seen at all.

Methodically, Eddie and I began a circle to the left around the base of the mountain. We were particularly anxious to locate two rooms of which we had heard reports. One was said to be directly beneath the talus mountain and to be more than a city block in length. The other was said to lead outward under the overhang and to proceed for more than five miles away from the Hole itself.

None of our party found either of these chambers. It is doubtful, to my mind, that the five-mile passage exists, although there may be a continuation of a room which can be reached only by swimming underwater in one of the lake rooms. This we did not attempt because of the lack of time.

The chambers which Eddie and I did enter, however, were many and beautiful. Almost every type of cave formation of which I have ever heard appears somewhere—if not profusely—in this subterranean fastness. Even the walls of the outer mountain room show rich and varied canopies of stalactites, flowstones, dripstone, "bacon rind," tiny helictites, and other forms. The coloring is mainly in the iron spectrum, ranging from delicate pastels to full, rich brown. There are also strange greens and formations of crystal white calcite.

Circling the main room, Eddie and I entered dozens of blind leads, mere wormways beneath the talus, leading nowhere. But always a drop away from the mountain revealed new wonders. Along the far perimeter of the lowest level we could reach, stands the Emerald Lake. The water is cool and crystal clear, taking its emerald hue from some source unknown to us. The water depth varies, roughly from five to thirty feet, and several overhangs were noted well beneath the surface. The running stream reported to be in the cave was not observed by our party. It may exist on a lower level than we reached, or in the great room under the mountain, if there is such a room.

Our trip around the base of the mountain was not too difficult. A great amount of climbing and some rockwork were necessary, but the way was safe enough, and there appears to be almost no scree on the slopes—most of the rocks being firmly wedged.

Under the northwest ledge we came upon a colony of bats. They were the only ones we saw in repose, and though there were a great many there, it did not seem to me that they were more than a fraction of the number we had observed flying the night before. (It is the opinion of the natives that the majority of the bat population does not reside beneath these outer ledges, but hides away in some unfound gallery.)

One of the most spectacular phenomena of the cave may be observed from directly beneath this northwesterly bat ledge. It is the westerly face of the mountain, which, from this vantage point, is seen to rise as a giant fang, sharp and forbidding, with an enamel of slick, glistening travertine encrusting it. This face of the mountain could not be scaled without ropes from above. It is as slick as ice and as formidable as the Matterhorn. We had to circle halfway back around the base of the mountain before we could climb back to the summit. By that time there was only about half an hour of daylight left. And we were anxious to reach the surface before dark fell and the bat flight began.

On the summit, we sat down to catch our breath after the exertion of the climb. We looked upward and for the first time began to consider the skyward journey. The prospect was anything but pleasing. Dimly, against the darkening sky, we could see a thread-like cable stretched across the mouth of the shaft. There was no sign of activity. We shouted and were told that it would take about ten minutes more to complete the rigging.

Following Eddie's suggestion, they were mooring one end of the cable securely to a peg in the ground on the opposite side of the Hole from the car. The bosun's chair was suspended from a pulley which rode free on this cable. Thus, the cable would run from the mooring, loop down the shaft, through the pulley, and up to the car. As the car advanced, it would pull the loop taut across the top of the Hole, lifting the chair up the center of the shaft instead of along the lower ledges. There would be no twirl. The safety rope was to be knotted loosely around the cable and used by the ground crew to pull the chair from the center to the edge, once the cable was made taut and the chair was hanging even with the surface. It sounds good. It probably even looked all right—but not from the bottom of the Hole!

It was getting cold. Eddie and I huddled together as close to the Coleman lantern as we could get. We discussed the cave. We tried not to think of the spider-like contraption above us. Ten minutes passed. Then twenty, and thirty. It grew dark. We became progressively more nervous. "The unexpected!" I complained. "It's always the unexpected that causes trouble. Who ever expected anybody to re-rig and then to test the gear on live subjects in the dark?"

There was some activity at the opening. Then a pebble came down with the zing and the splat of a rifle bullet. We yelled for care topside and were assured that all would be well in a matter of minutes. Then we saw the half-inch manila line wagging back and forth at the edge of the Hole as they tried to stretch it across. This is a difficult task, for it must be thrown clear of bushes and small trees. But from our position on the bottom, their efforts appeared to be the confused fumbling of idiots. "Why don't they get started?" Eddie muttered. I was too disgusted to reply.

Finally, just as the rope appeared to be safely across, one end came loose, and the hundred-foot length of it, held only at one end, fell down the shaft. It seemed to come slowly, straightening itself out leisurely like a giant, overfed snake. Then it popped straight, with a sound like a long-drawn-out whiplash. There were curses from above, and the rope began to be hauled upward.

Eddie and I made no comment. Unreasonably, our confidence in our teammates was evaporating at the prospect of being drawn up the center of the dark shaft with such rigging. "What have you got that cable moored to?" I howled up to the surface. Ralph, as urbane a person as I have ever known, lay down and poked his head over the edge to mollify us.

"It's moored to an iron peg in the rock," he called patiently.

"Only one peg?" I yelled back.

"It's perfectly secure," Ralph assured me and began to go into a lengthy discussion of the tests to which he had subjected it.

"Never mind," I interrupted. "Let's get going. Only make sure you don't drive the car an inch too far and pull out the peg while we're hanging at the top." With as much calmness as can be commanded by a man shrieking at the top of his lungs, Ralph reassured us further. Then he disappeared from view.

The bats came out. First just a few. Then so thick that a steady rain of guano was hitting our helmets. I quit looking upward after receiving several eyefuls. The air was beginning to sound like a roaring torrent. It was pitch-dark.

People react differently in their subconscious efforts to protect themselves from the effects of nervousness. I have seen many strange habit patterns which are simply sublimations of the fear instinct: Mine is anger. I cursed away mightily for another twenty minutes before the swing came down.

Then I cursed ten minutes more while suspended sixty feet off the floor; the rigging was fouled with the safety rope. Finally I began going up again. The ascent was perfectly smooth and effortless. The sound of the cable whispering through the pulley was reassuringly steady. But now I was in the unbelievable vortex of bats. Contrary to their custom, several thumped me on the legs.

About ten feet from the top, I began yelling to my companions to slow the car and guard against pulling up the stationary end of the cable. They were as concerned about this, or almost as concerned, as I. We negotiated the last ten feet at a snail's pace. And I was sweating profusely in spite of the cold.

I refused to look down. I kept looking up, in spite of the guano, and there, through the bats, were the Seven Sisters of the Pleiades directly overhead.

Ralph let go the safety rope. Its big open knot slid down the cable with a "swoosh" and fetched up against the pulley; its free end struck the side of my head with stunning force, but I clung to the bosun's sling. It is an eerie feeling to hold on to something which you know will be the one thing that falls if anything does. "For God's sake, pull this damned thing in!" I demanded.

They pulled. But I did not move very far. The sag in the cable was still too great. Floyd had to run eight hundred feet down the road to the car and inch it even further ahead, applying more pressure to the mooring. Then he came back, and they brought me over against the ledge, but there I stuck. All of their work to re-rig was useless, insofar as those last three feet were concerned. The cable sag placed me in exactly the same predicament that Floyd and Ralph had been in that afternoon.

If anything, the situation was worse. I was slightly lower and involved with an extra rope in which I nearly hanged myself. All in all, it took fifteen minutes to get me over that ledge. At that, I had to repeat Floyd's maneuver of standing upright in the swinging bosun's chair.

I lay on the ground for a moment and then, as Ralph had done, walked away from the Hole. I did not want to look at it or go near it. It was then that I had my most difficult chore of that nerve-racking day. I had promised Eddie as I left the cave floor that I would check the cable mooring before he came up. This was easy enough to do, but it meant returning to the edge of the pit, lying down, and sticking my head back out over the side. For some reason it was more difficult to return to the edge of the pit than it had been to get into the swing and make the ascent. As soon as I reported to Eddie, I got back away from the Hole again. But by the time the chair had been lowered to the floor for him, my short-lived repugnance for the Hole had vanished, and I went back to the edge to help get him up.

It was the same story: A clear run upward through the bats. The stalled swing over the middle of the Hole, in the last few feet. The hand-over-handing of the pulley over to the ledge. And, finally, the nightmarish minutes of struggle to get him across the ledge. By this time, it was dark, and the column of bats had reached its full height.

Comedy of unpreparedness though it was, this initial expedition satisfied us as a preliminary survey. Considering the size of the Hole and the time we allowed, we had not done badly. But all of us had had our fill of this particular rigging. As we drove homeward through the cold, clear night, we determined to come back the following weekend. And we did, but with far safer and more elaborate equipment.

Cave of the Vampires

As is well known, the expression "vampire" originally had nothing to do with bats or caves. In Slavic countries the vampire of folklore was a walking corpse, which left its grave at night to suck the blood of sleeping victims.

Only after the discovery of America did it become known that there were living creatures with habits resembling those of the legendary vampires. Columbus and the Spanish Conquistadores learned about bats that attacked animals, and it is not surprising that the several species of sanguivorous bats should come to be known as Vampire Bats.

The paucity of information about the northward extent of their range in Mexico, and the associated public health problems, led the author in 1946 to organize a party of six members of the National Speleological Society to search for vampires. This chapter tells the story of a second, follow-up trip, the same year.

CHARLES E. MOHR

The spot where the cart trail turned off the Pan-American Highway was in the lowlands, four miles north of Valles and eighty miles inland from the port of Tampico. As I told Dan and Jule, it didn't look any smoother than when I had been here in June, ten weeks before. Fifteen miles per hour was top speed. Here and there, we had to jockey the heavily loaded station wagon over irregular rock ledges, first filling in the low spots with flat stones. There was a washout to bridge, and muddy spots to fill with cactus and mesquite.

The little group of a dozen thatched huts surrounded by sparse trees and scrubby fields of weeds was Los Sabinos. Here we'd find our guide and switch to horses for the three-mile trip to the cave.

Miguel Herrera—"Mike" to us—was among the group sitting quietly in the shade of a roofed structure, open like a carport.

"*Buenos días, Señor,*" he called. "*Murciélagos?*" Was I looking for bats again?

"*Sí, Mike, Vampiros.*"

Jule, who spoke much better Spanish than I, immediately began to ask all about the Vampire Bats. Did the bats come to the village every night? Weren't the Indians afraid of them? What happened to the livestock that was bitten?

The villagers knew all about the vampires and discussed them as calmly as American ranchers talk about coyotes. Yes, sometimes on a moonlit night they saw the vampires, knew that often they swooped down on their cattle and burros, attacked their goats and even their chickens. No, the large domestic animals seemed none the worse for the blood-letting, but chickens and turkeys often died from loss of blood. Yes, the Indians did take precautions; they brought the fowl into the houses or put them in screened or tightly woven pens.

"What about people? Do the vampires ever bite people?"

Always they insisted that only animals were bitten, never people. Yet I knew that in many parts of the tropics, in northwestern and central Mexico, for example, people had been bitten; some had even died from vampire bites—the bites of rabid animals. In this eastern section of Mexico, though, no hydrophobia cases had been reported. Bites by healthy vampires sometimes leave such slight wounds that they pass unnoticed. Often, however, they leave an unmistakable oval wound. Children who receive such wounds are said to have been bitten by *brujas,* or witches.

"Couldn't a person bleed to death?" Jule asked. "I've read that the bats may bite into an artery or large vein."

"You've been reading the wrong books," I replied, laughing. "The wound is quite shallow, and the loss of blood is seldom serious. A complete meal for a vampire might amount to a few tablespoonfuls. You cheerfully give a pint of your blood to the Red Cross several times a year. I'm sure you would never miss the little that a vampire took."

Jule repressed a shudder, but she was no less eager to visit the Cave of the Vampires.

A curious series of events had led to this odd scene—Jule Mannix, Powers model, lecturer, radio and TV performer, back scarcely a week from a Hollywood screen test, now garbed in dungarees and T shirt, clamoring to penetrate the jungle and a cave to come face to face with Vampire Bats.

It was in the halls of the Academy of Natural Sciences of Philadelphia, oldest science museum in America, that the expedition had been conceived. Just back from leading a six-man expedition to a score of Mexican caves, I stepped out of my office one day and saw Dan Mannix walking through the museum lobby. Dan, a lanky six-feet-four, has taken turns at being a sword-swallower and fire-eater, falconer, photographer, and writer.

"I stayed at Casa Humboldt for a couple of days last month," was my greeting to him. One of the best-known houses in Taxco, the place had served as headquarters for Alexander von Humboldt, celebrated German naturalist, at the start of the nineteenth century. Dan and Jule had stayed there for six months while they hunted iguanas with their trained Bald Eagle, Aguila, in the hills above the picturesque town. They had made a spectacular motion picture on their Mexican adventures, and I had launched them on a lecture tour by putting them on the academy program immediately after their return.

"What were you doing in Taxco?" Dan wanted to know.

"Exploring caves. Hunting blindfish, insects, Fruit Bats, Vampire Bats. . . ."

"Vampire Bats!" Dan interrupted. "Did you bring any back with you? I've always wanted to see a live vampire."

"I collected half a dozen," I told him. "But they're museum specimens now."

Dan was visibly disappointed, but a few minutes later, in my office, he studied the specimens intently. Just where did I get them? and were there many more? he wanted to know.

I had found vampires in only five out of some twenty caves we explored, I told him, but if he wanted to visit these five the next time he was in Mexico, I'd be glad to give him directions for finding them.

Dan left, but a week later he was back with a proposition.

"Jule and I are driving down to Taxco for a couple of months while I do a story. How about coming along and leading us to the vampires? I'm sure my editor would take an article on the trip. You could check up on all the vampire colonies in northern Mexico, hunt other cave animals, get more specimens for the museum, take lots of pictures. . . ."

Dan went on to describe the advantages of my going to Mexico with them. He was so persuasive that the museum officials approved the trip—my second of the summer. Jule had hurried home from Hollywood. She refused to be left behind. Joe, her fourteen-year-old brother, and Wriggles, their Cairn puppy, rounded out the party.

I had briefed them pretty thoroughly on what equipment we would need. Besides the four cameras, we had a mountain of camera cases, tripods, and cartons of flash bulbs, a huge coil of one-inch hemp rope, boots, hard hats, flashlights, gasoline lanterns, carbide lights, cans of carbide, heavy leather gloves for everyone, canteens for water, emergency rations, DDT, a well-stocked first-aid kit, antivenin, and a snakebite kit. Since I would be taking back certain specimens to the museum, I had cages and bottles for live animals and preserved ones, as well as equipment and supplies for skinning and mounting small mammals, especially bats.

Valles was our first important stop, and as soon as we had established ourselves at a convenient motel, and deposited Wriggles, we set out for Los Sabinos and the Cave of the Vampires. Mike joined us, with his cousin, Malaquinas Herrera, leading two horses. Jule, an accomplished rider, took one, and Mike rode the other, carrying the heaviest equipment. The rest of us followed on foot.

A well-worn foot trail skirted several fields, then followed the edge of dense woods filled with a tangle of palmettos, scrub trees, and thorny vines. Abruptly it turned into the woods. The riders had to dismount, for the vegetation had closed in on the trail since June. The horses were left tethered to a tree. As I looked into the jungle, I doubted whether I could have found my way from memory, had I been alone.

Fifteen minutes later, the land fell away sharply in front of us. We were at the brink of a heavily shaded canyon. Turning to the right, we picked our way along the narrow trail, as it descended steeply to the floor of the canyon. There we doubled back and went more gradually down hill. It became noticeably darker. The canyon was roofed over. Even behind us, the light was too dim for snapshots. As our eyes became adjusted, we began to appreciate the wild beauty

Cave of the Vampires

Down, just within the entrance of Cave of the Vampires, Los Sabinos, may be seen the limestone partition which divides the cave. *Charles E. Mohr*

of the jungle scene. Before us, we could make out the eighty-foot entrance to the cave.

Somewhere above us, along the canyon rim, a tremendous chorus arose, "Cha-cha-láca, cha-cha-láca, cha-cha-láca." When I had first heard this call, earlier in the summer, I had recognized it instantly as the deafening song of the pheasant-sized chachalaca, which provided the noisy climax to one of Arthur Allen's famous bird recordings. There was no mistaking it. Other exotic sounds beat upon our ears. A flock of large, green parrots flew swiftly past us into the cave, their raucous cries echoing from the dimly lit corridors.

A colossal limestone partition divided the entrance. To the left yawned a black abyss. To the right, at the bottom of a great rockslide, we could see a level floor leading off through a weird setting of eroded windows and narrow portals.

We reached for our lamps. The Indians soon had the Coleman

lanterns going. Jule and Dan snapped their electric miner's lamps to their helmets and the six-pound, metal-encased batteries to their belts. I filled my carbide headlamp with an ounce of carbide, added an ounce of water, waited a moment for gas to be generated, cupped my hand over the reflector, and spun the corrugated steel wheel against the tiny flint. Sparks flew, and the gas ignited with a distinct pop. The flame was an inch long, so I reduced the flow of water, cutting the flame to an economic half-inch length.

The Indians obviously hadn't much confidence in our lighting equipment. They each carried a big pocketful of rough, home-dipped candles that we had seen hanging from the ceilings of their huts. Even when I showed them what a powerful beam our flashlights produced, they weren't impressed. Their candles couldn't go dead the way batteries do, they said, and bumping or dropping one doesn't put it out of commission like a fragile gasoline lantern. And they were right. Even the rugged carbide lights are temperamental. Many a time I have relied on candles to get me out of a rough cave, when all the modern lights have failed. Stowed away in our own packs were thick plumbers' candles and matches in waterproof boxes.

"Let's go, everybody," I shouted. "This is it, the Cave of the Vampires." We set out in single file, moving along easily in the narrow corridor. Here and there we could look through windows into the black void to our left. It was a balcony we were traveling along, then a ramp that led down to a spot from which we lowered ourselves into the big room.

"It's as big as Grand Central Station," cried Jule, as we probed the ceiling with the flashlights. The cave seemed to end here, but at floor level there were many black pockets, entrances to crawlway passages. Most of the passageways went in only a short distance and stopped. One, I remembered, led to a bat chamber, but which one?

Mike and Malaquinas stopped in front of a hole that seemed no different from any of the others. They got down on their stomachs and started to wiggle into it.

"This is a pretty tight squeeze, and you're not dressed for crawling," I told Jule. "Why don't you wait out here for us?"

"I should say not," she retorted, and began crawling after the guides, with Dan and me right behind her. Almost immediately her hard hat started sliding down over her face. It didn't fit snugly enough over her hair. She took it off and pushed it ahead of her as she crawled. The passage got a little bigger, but when we tried to travel on hands and knees we banged our heads sharply on the ceiling. There was no way to go on except by sliding along on our stomachs.

"For the first time in my life, I felt panicky," Jule told me afterward. "I guess I'm subject to claustrophobia after all. It was like crawling through a rough sewer pipe. The sides seemed to be closing in on me. I could hardly breathe. I was sure the Indians would get stuck, or come to a dead end, and we'd have to try to back out of the tunnel."

Just when she felt she must scream, the Indian ahead got to his knees, then stood up. Shakily, she followed and put on her helmet. How wonderful to be able to stand up. Soon everyone was in the room.

"Where are the bats?" she asked. Mounds of rice-like, dark-gray "scats" covered the floor—the guano of insect-eating bats. But nowhere did we see a bat. They'd abandoned this roost, possibly just for the summer, I decided. This wasn't the important bat room anyway.

"We don't have to go back the same way?" Jule protested. I nodded. It was the only way out.

So back we crawled through the tunnel, and out into the big room, which now seemed even more tremendous than before. I knew that the corridor to the vampire roost lay off to the right.

Halfway around the room we found an inconspicuous, low opening that led downward behind a rock partition into a long, high corridor. At the entrance, Mike lit a candle and left it on a rock. Every hundred yards he left another lighted candle. As he set up the last one, we saw that our guides were standing on the brink of a deep pit. One by one, we inched over and joined them on the ledge. We played our flashlights back and forth on an overhang about fifteen feet below us. It projected out over nothingness.

This was why we had brought the ropes. Mike scrambled down to the overhang, doubled the one-inch rope, and anchored it around a stalagmite. The rope was 150 feet long; doubled, and looped around the stalagmite, it would reach about seventy feet. Then, without warning, Mike scrambled over the edge and disappeared. I slid down to the overhang and peered over. I could see his wide-brimmed hat

Brink of the deep pit leading into the Bat Room, where the vampires lived. *Charles E. Mohr*

far below and receding rapidly. Just as he reached the end of the rope, the shaft began to level off. He didn't need the rope for the rest of the descent.

Dan, Malaquinas, and Joe formed a human chain, handing down to me the heavier equipment, which we had piled up on the ledge. With a light line we carried for that purpose, I lowered it piece by piece.

Then, grasping the heavy rope, I climbed down hand over hand, taking advantage of occasional toeholds in the rocks. As soon as I was safely out of the way of possible falling rocks, I called Joe. He came down quickly and went on down the slope to join Mike.

It was Jule's turn now, and, taking a firm grip on the rope, she slowly inched backward over the edge into the near-darkness. She was almost down when suddenly there was a report like a rifle shot. Then another. Like explosions, they came echoing up the pit from the room below.

"Cut it out, Joe," I cried. He had become bored waiting for the others to come down and had thrown two of Dan's big flash bulbs against the wall. He had forgotten for a moment that we had agreed not to clutter up the cave with our debris. With bulbs, there was danger someone might get cut.

He didn't throw any more, but the sudden noise had disturbed a group of bats, and now they erupted from the pit, up past the rest of us. We could hear their shrill squeaks, feel the swish of their ghost-like wings as they barely missed colliding with the climbers.

Our first thought was for Jule. She had stopped. Startled by the explosions and now caught in an inky blizzard of bats, she had frozen to the rope. Fortunately she had found a foothold in the rock, so she wasn't dangling unsupported.

Dan shone his light down from above. "Are you all right, Jule?" he asked anxiously.

"Are these the vampires, Charles?" she screamed. "They won't attack us, will they?"

"Not unless they're rabid," I started to tease her, but realizing that she was really frightened, I hastily added, "These aren't vampires, Jule. They're little insect-eating bats, and they won't bother you. You can come down now."

When everyone had reached bottom, we pushed on into the bat room. Sixty feet wide and ten to fifteen feet high, it had a shallow lake at one side, a terraced cascade on the other. The water was perfectly clear, and kneeling down at the edge of the pool, I saw small, whitish fish that looked just like the famous blindfish I had seen in another cave, in central Mexico.

"I've found some bats," cried Joe excitedly from one of the alcoves. "Bring your net."

We all converged on Joe. Just above him were half a dozen large bats. I swung the long-handled insect net, just as they took wing. All but one escaped, but that one was really a handful. Carefully, but somewhat clumsily, with my leather gloves on, I extricated the snarling creature from the net and held it up by its wing tips. It stretched the full length of my hand and forearm, about eighteen inches.

"It's got a baby!" exclaimed Jule.

We focused our lights upon it and discovered a tiny bat, covered with fuzzy down, clinging to its mother. It couldn't have been more

Adult Mexican Fruit Bat, *Artibeus*. The formidable teeth are used for splitting the solid Mexican fruits and sometimes for attacking and eating smaller bats. *Charles E. Mohr*

than a week old and was completely helpless. It was partly hidden by the scooped-up spread of the tail membrane, which became a kind of hammock as the mother bent her legs up toward her body. It had no bony tail such as northern bats have.

This was no vampire. I pointed to the spear-like projection that rose from the front of its long nose. The baby also had a prominent spear on its nose. These were Fruit Bats; vampires lack the spear. As the big bat struggled to escape, it bared the most formidable set of teeth we'd ever seen in a creature of its size.

"Why does it need such big, sharp teeth, if it lives on soft tropical fruit?" Dan asked.

"Some of the fruits are pretty solid," I explained, "and at times these bats are carnivorous, attacking and eating smaller bats."

"Señor, look here!" called Mike. He pointed to a strange forest of extremely thin shoots. They were seedlings, pure white and as much as two feet tall. Wherever we found the little white groves we noticed above them dark spots on the ceiling where Fruit Bats had been roosting.

We found dozens of the partly decayed hulls of the seeds from which the attenuated "trees" had sprouted. These seeds represented the undigested remains of the bats' meals of fruit.

"How can they grow in the dark?" Jule wanted to know.

"Light isn't necessary for germination," I told her. "There is enough stored-up food in these seeds to give the young plants a good start. They may also absorb some nourishment from the rich guano they are growing in. But they are doomed, because chlorophyll cannot develop where there is no light."

Suddenly I heard bats flying. I knew from the strong, loud wing-beats that they were not the little insect-eaters. They were Vampire Bats! They were flying in one of the corridors that paralleled the main room.

I ducked under a stalactite barricade, with everyone right behind me, and stopped short. On the floor was a series of irregular pools of evil-smelling, brownish-black, tar-like liquid. Directly over the strange pools were half a dozen bats—vampires!

Trying to keep out of the black pools, I reached up awkwardly with the insect net. Instead of flying, these vampires dodged nimbly, scuttling backwards and sideways like crabs or spiders. They had no trouble avoiding the extended net, and when I maneuvered into a better position, they darted into a crevice where I could not reach them. I knew there were others, so I wasn't too much concerned by their escape.

"Over here," called Dan. He had his big press camera in one hand and was fumbling with a flash bulb. Annoyed, he pulled off his bulky leather gloves.

In a deep, bowl-like depression, just at eye level, were three half-grown vampires. The beam from my flashlight cast elongated shadows which accentuated their strange appearance. While other species of bats sometimes slither along a ledge, they do so with their bodies flat against the rock, and they pull themselves along with tiny thumbs, which project from the elbow-like wrist joint.

Not so the vampires. They stood high on all fours—on their long hind legs and their tremendous thumbs. No wonder they could dodge about so fast, even bounce around like rubber balls. They looked like some strange race of little men as they peered out at us. Jule said they didn't look like real animals, they looked like goblins. "Evil little men" is the translation of one of the Indian names for them.

Dan brought his camera within a foot of the three vampires. Their escape cut off, they huddled into the deepest part of the depression, their little bulldog faces and twitching ears forming an amusing picture.

As the camera bore down upon them, one of the vampires somehow slid himself behind the other two, leaving them to face the menacing lens. A moment later the flash went off, and one of the

The three young vampires just before they started their game of musical chairs. *Charles E. Mohr*

forward pair dived for the rear and shoved his hiding brother into the limelight again. But only for a moment did they stand staring into the camera. The scramble was on again, and a new member of the trio was in hiding. This vampire version of musical chairs went on for minutes until one of the guides called that they had found "hundreds" of vampires.

They had. And more of the smelly black pools covered the floor. They were, of course, the excrement of Vampire Bats. Completely different from the dark-gray, rice-like, chitinous guano of insectivorous bats, or the seed-peppered guano of the fruit-eaters, this tar-like material was what remained from their sanguineous fare. Neither fragments of insects nor seeds go through a vampire's alimentary canal, only blood. And with "predigested" food of this sort, vampires don't need a long, complicated digestive system as other creatures do. Virtually a straight tube will suffice. The fluid portion is speedily eliminated, and the solids are absorbed easily.

The vampires were hanging in groups in several alcoves. They looked more furry than most bats, like miniature, reddish-brown bearskin rugs. Hanging head downward, they watched us, their partly open, grinning mouths revealing only a few teeth—but what teeth they were!

Dan started to reach for one bare-handed, then thoughtfully slid his hands back into the heavy leather gloves he had removed. With a swift motion he grabbed one of the bats. It struggled savagely. Noticeably smaller than the Fruit Bat, the vampire still had a wingspread of thirteen inches, measured four inches from its toes to its nose. It was a conspicuously fat bat, compared to other species. No wonder the Vampire Bat had been named *Desmodus rotundus*. Its body was almost as thick through the middle as it was long.

"Help me put him in the cage, Jule," Dan said. Jule hesitated for a moment, then held up the small metal cage. She lifted the sliding door and Dan started to stuff the bat in. As its head approached the cage, the bat seemed to lunge at Jule's gloved hand and nod vigorously. A moment later, he spit out a little strip of brown material the same color as the glove. Afterward Jule discovered that an oval groove a quarter-inch long had been neatly sliced out of the back of her glove by the vampire's two razor-sharp upper incisors. The bat's set-back canine teeth are even more murderous-looking but evidently are not used in their normal blood-letting activities.

Examined at close range, the Vampire Bat has a face one doesn't forget in a hurry. Most bats have fairly pointed faces, but the vampire's is almost flat. Its nose is broadened and flattened, so as not to interfere when the bat uses its incisors to make a razor-like sweep into the victim's skin.

We moved about the room, trying to get an idea of the size of the population. Some of the vampires scuttled off across the ceiling, but quite a few launched themselves into the air. We estimated that there were at least five hundred vampires and perhaps as many insect-eat-

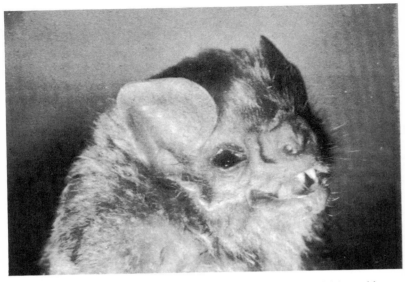

Close-up of Vampire Bat showing sharp incisor teeth which enable it to slash victims to draw blood. The blunt nose enables this bat to bite its prey without otherwise disturbing the victim. *Charles E. Mohr*

ers and Fruit Bats. There must have been several hundred in the air, flying in and out of the alcoves and passageways, which continued for more than 150 yards. The small bats flew mostly in circles; the vampires headed off in a straight line and disappeared.

So great was the congestion of the flying mammals that even their sonar type of navigation didn't prevent many wing brushes and a couple of outright collisions. Several times bats flew against us, but there was nothing suggesting the aggressive, head-on attack of a rabid bat, such as I once witnessed in southern Mexico.

A sharp exclamation in Spanish was followed by a crash of metal and breaking glass. Mike had stepped back into a pool of slippery vampire guano, losing his balance. Trying to catch himself he had banged the fragile lantern against the wall. One less light, I figured, but we still had plenty. My carbide headlamp was sputtering though, so I stopped to probe the nozzle with cleaning wire. It didn't help. I used my hand flashlight instead. The batteries in Joe's light were getting weak. We'd better be heading toward the surface, I announced.

But first I wanted some blindfish for Dr. Charles Breder, at the American Museum of Natural History in New York City. Ichthyologists suspected that each cave here has its own species or subspecies of this interesting cave fish. They were eager to get some live ones from each of the five caves from which they were known.

The fish were easy to catch. All we had to do was toss a pebble into the pool, and they rushed over to investigate. Increased sensitivity to vibrations probably compensates for lack of sight. And since most disturbances are associated with falling insects, guano, or dead bats, the response of the fish is swift. We scooped up three fish and transferred them to a thermos jug. If we had collected more, there might not have been enough oxygen for them, and all might have perished.

Meanwhile Dan and Jule had captured several more vampires. Eager to observe their feeding behavior, we planned to take them back to the motel. I wanted a few specimens for the museum collection, so I called Malaquinas, who carried a .410-gauge shotgun, and pointed to a cluster of a dozen or so on the ceiling. They were out of reach of the net. Raising the gun deliberately, he fired once at the vampires. The report was terrific.

Every one of them fell, and both guides, Joe, and I dashed forward. We all wore gloves, but even so we reached cautiously toward the fallen bats. At that moment they started flying off, one after another, until only one was left. This one, I saw, had been shot through the head and was dead.

The shot had caused almost all the other bats in the room to fly off. I told Malaquinas to shoot again, at one of the few remaining clusters. Again, bats fell like rain, and this time we collected three. The rest leaped into the air as we approached to pick them up. Vampire Bats are the toughest of bats. Only a pellet through the heart or head will stop them.

"Let's go now," I repeated. No urging was needed. All of our lights were weak. The still burning candles were a welcome sight as we reached the foot of the shaft and started the slow ascent to the surface. The guides scampered up the rope first, then hauled Jule up, and afterward Joe and Dan and me.

Back in town we examined our specimens. Five hungry vampires. What was there to feed them? We bought a live chicken in the market, killed it, and collected the blood in a saucer. Unfortunately it coagulated faster than we expected, and by the time the bats discovered it they couldn't lap it up.

It is possible to neutralize the clotting agent in blood, and once defibrinated, the blood can be kept for a considerable time if refrigerated. But we hadn't brought the necessary chemicals and probably wouldn't be able to secure them until we reached Mexico City. I remembered, though, that Dr. Alfredo Téllez Girón always kept a supply of liquid blood in his laboratory in Mexico City. He was head of the Department of Agriculture's Animal Pathology Laboratory.

"It's just about three hundred miles. We could make it easily tomorow if we started early," I said.

Most of the night we spent getting our equipment in order and packed into the station wagon. We were on the road well before dawn. Breakfast in Tamasunchale, a winding, hair-raising ascent to 7,800 feet through the breath-taking mountains beyond, and a dash across the high desert. We were in Dr. Girón's laboratory in mid-afternoon. He was cordial and cooperative.

"Vampires don't live well at this altitude," he said, referring to Mexico City's 7,400-foot elevation. "We've had much more success in keeping them alive at our Acapulco laboratory, at sea level."

The bats didn't look as though they were likely to survive the night. Dan tried unsuccessfully to feed them with an eyedropper.

"Give that to me," Jule said finally, after watching Dan spill blood over himself, the bats, and the table. "You never did know how to feed animals."

Putting on the leather gloves, she picked up one of the bats, spoke to it, scratched its head, and finally persuaded it to feed. It licked the blood off the end of the eyedropper.

"You see, they aren't really mean. They were just frightened and upset." They did seem nervous, more so than any bats I'd ever seen.

Within a few days the vampires fed calmly as Jule squeezed blood from the eyedropper, a drop at a time. We had moved down to Taxco, at 5,000 feet. Here I believed the bats would live, because I found several vampires in a cave near town. Occasionally they have been found at even higher elevations, but they are more abundant in the lowland tropics.

Our vampires got so tame that they would come bouncing out of the cage when the door was opened and go scuttling around in search of their handout. Jule even allowed them to climb up on her dress.

"I never became really fond of them," she said afterwards. "But I got so used to them I didn't mind their looks. They're really handsome in their way . . . like little gargoyles."

After a while, Jule found it more convenient to put the blood in a dish and let the vampires feed themselves. Now it was easy to see that instead of putting their mouths to the liquid and sucking it up, as they were long believed to do, the bats lapped it with their darting, long, pink tongues. The bats gorged themselves until their bodies were as round as ping-pong balls. Then they would fly to the top of their screened cage, hang upside down, and preen themselves as a cat might do.

In one of the laboratories we saw a captive vampire land close to a big, white rooster and cautiously follow it around. Finally it made a quick pass at the rooster's leg, biting it in the heel. When the bird began to bleed, the vampire started lapping the wound. The rooster seemed completely unconcerned, but we knew that a few attacks like this would be fatal.

Other victims, like a burro attacked by another captive vampire, seemed equally undisturbed by the uninvited guest. Perhaps the victims knew from repeated visits of the vampires that it was impossible to dislodge or discourage them. As a rule the victim didn't flinch when bitten. But younger bats were obviously inexpert. Their slashing seemed largely a matter of trial and error, and their prey often exhibited signs of pain.

The vampires evidently learn where capillaries are abundant, close to the skin, where a quick, shaving slice will painlessly remove enough skin to provide a steady flow of blood. Dan Mannix claimed that the older vampires practice both surgery and psychology in their assaults. They learn which areas have few nerves and lots of blood, he said. After stealthily stalking their quarry, they gently nip an ear, the tip of the nose, or the underside of a toe. If the victim stirs, they wait patiently until their quarry has gone back to sleep. Then they steady themselves "like surgeons preparing to make an incision with a scalpel," as Dan expressed it. Opening their mouths to the fullest, the vampires make a swift, downward slash with their exposed incisors. If, as is generally true, the quarry is unaware of the attack, the bats begin to lap the flowing blood and continue until they are gorged.

Does the wound continue to bleed? someone always asks. Occasionally it does. But the excessive bleeding may be due to a clumsy bite or a deeper one than normal. Certainly there is little evidence to suggest that an anti-coagulant is introduced into the wound from the attacker's saliva.

Dr. A. G. Ruthven, former president of the University of Michigan and a zoologist of note, recorded that one night in Colombia all the members of his group were attacked by vampires and that in the morning "the whole party was covered with blood from the many bites of these bats." Yet the men had not been disturbed in their sleep. There was no pain involved in the bite, he declared. Medical records in Trinidad show that less than one third of the people bitten were actually awakened by the attack.

In eastern Mexico, in Vera Cruz, and in San Luis Potosí, where we were, attacks really did seem to be limited to domestic animals. Walter Dalquest, during several years of collecting mammal specimens for museums, declared that wild animals never show the telltale scars that are so common on cattle, horses, and burros. And even though his party camped for several months near caves known to have vampires, none of the men was ever bitten. Yet two horses roaming near a bat cave became so emaciated in about two weeks that they would no doubt have died if they had not been moved.

Dalquest believes that vampires were uncommon, possibly rare, in eastern Mexico in prehistoric times. But following the conquest of the country by Europeans, the vampires flourished because of the increased food supply in the form of the relatively helpless burros, horses, and cattle. As a result, the vampire has increased in numbers until it is one of the most common and widespread mammals in eastern Mexico.

To determine, if possible, whether the vampires reached the United States, I had already made several trips to Texas and Mexico. Search for them in the numerous caves in the Edwards Plateau, north of San Antonio, had proved fruitless. Total absence of the telltale vampire pools was convincing evidence that no bats were present.

Fossil remains of vampires, perhaps a million years old, had been found in San Jacinto Cavern at the edge of the desert north of Monterrey. I visited Cueva del Carrizal, near the famous Golondrinas Mines and the spectacular mountain, El Candela. This, too, was at the edge of the desert. Two of the workmen at the mine claimed that *vampiros* lived this far north, but no evidence of them could be found.

We investigated every cave we learned about, including spectacular Cueva la Boca, high on the side of a peak about a mile east of the highway at Villa de Santiago. No vampires did we find here, or in the almost equally huge Cueva del Abra, overlooking the highway, a ten-minute climb above kilometer post 523.

But at Cueva del Pachón, just north of Antiguo Moreles, there were vampires. This, the northernmost colony we found, was just 250 miles south of the border. These cave vampires and a few found in the jungle near the coast, 175 miles from the Río Grande, represent the Vampire Bat's nearest approach to the United States.

Will vampires eventually be found in our Southwest? I doubt it. The arid lands of extreme northern Mexico support few domestic animals. Caves are few and scattered and are easy to check for possible vampire occupancy. The bats do, of course, roost in hollow trees and in the walls of stone buildings, but probably 90 per cent of the known colonies live in caves. The coolest, dampest caves, we learned, were the ones picked by the vampires. They shunned the big caves with large entrances. Since there is no evidence that vampires are migratory, and since the possibility of their gradual spread northward is so limited by hostile terrain, the danger seems remote.

"Danger" is the proper word to use. In more tropical portions of America, the Vampire Bat represents a serious health and economic problem. Not only is this bat at times the carrier of paralytic rabies, responsible for scores of deaths during outbreaks, in Trinidad, for instance, but it also has been convicted of spreading murrina, a major plague of the livestock industry in Mexico.

Derriengue, the Spanish word for the disease, means "to break the spine." This highly fatal illness has wiped out whole herds of cattle in southwestern Mexico. Long a mysterious disease, murrina, or *derriengue,* was thought to be caused by a protozoan parasite, a trypanosome, transmitted by vampires when they bit the cattle.

After years of research, supported by the Rockefeller Foundation and the Mexican government, Dr. Harald N. Johnson and Dr. Girón found rabies virus in the salivary glands and brain of a paralyzed cow and in similar organs of Vampire Bats captured in a nearby cave. They concluded after further investigations that the mysterious *derriengue* was actually rabies.

When four members of one Mexican family in the State of Sinaloa died from the bites of rabid vampires, specialists from Mexico City rushed to the scene and attacked the bat colonies with flame-throwers, burning them out of caves, trees, and vacant dwellings. Equally drastic steps have been taken at other focal points where vampires were known or suspected to be rabid.

Experts of the United States Public Health Service and the various state departments of health are making periodic checks of the major bat colonies, often with the assistance of the speleologists who first explored the caves and discovered the existence of the bat concentrations. The importance attached to this project lies not in the likelihood of an invasion of vampires, but in the established fact that rabies has occasionally been found among fruit-eating and insect-eating bats. Two nonfatal attacks on human beings in the United States in 1953 were reported by the Public Health Service. Since both bats were captured and identified as belonging to migratory species, it is possible that they had been bitten and infected by rabid vampires during their winter stay in Mexico. Bat banding has proved that *Tadarida,* which spends the summer in the Southwest, may fly far into Mexico each autumn, into Vampire Bat territory.

How the rabies virus reached the vampires in the first place is a puzzle, for bats seldom roost where a dog, fox, raccoon, or skunk—all of which may carry rabies—might find and bite them.

Whatever the original source, the virus in bats varies somewhat from the classical canine rabies strain. The virus may have changed its properties still further when it was transferred to non-sanguivorous hosts, to insectivorous and fruit-eating bats. At any rate, current investigations by the Public Health Service and the Army Medical Service indicate that the bat virus found in Mexico is less virulent, possibly nonfatal.

"In any event," concludes Walter Dalquest, the vampires are "a menace to livestock and to the health of man in . . . Mexico at the present time and may be expected to be increasingly so in the future. Control of the bats promises to be a problem of major importance, and effective control may be impossible."

Assault on Schoolhouse

West Virginia, known for the rugged character of many of its caves, boasts of none more difficult than Schoolhouse Cave.

Ida V. Sawtelle, veteran of many cave expeditions, operates a dog training and boarding establishment in Brooklyn, New York. That is, she operates it when she is not visiting caves, for as this is written she is on an extended cave exploring trip in Europe.

Tom Culverwell's drawings and map of Schoolhouse provide a realistic impression of the obstacles to be overcome and enable the reader to follow the explorers' progress through the cave. Culverwell was one of the first to conquer Schoolhouse.

IDA V. SAWTELLE

The name "schoolhouse" may conjure up, for some people, a picture of simple childhood days in a one-room school, but to the spelunking fraternity it means the roughest, toughest, orneriest hunk of matter that ever was carved into a cave. One writer has called an assault on Schoolhouse Cave a postgraduate course in underground exploration, and so it is. Although it has little to offer in the way of beautiful formations, it has the hardest, sheerest, and most breathtaking climbs of all the caves in North America. One life has been lost in this cavern, but still the nation's best climbers risk their necks to search its inner depths.

One hot Fourth of July, I joined twelve other spelunkers in a West Virginia corn field. We prepared a hearty breakfast on our gasoline camp stoves before leaving the daylight for an arduous stint of cave-climbing in the darkness of Schoolhouse.

Joyce, our camp cook, dished up double rashers of country ham and eggs, hunks of dark bread and jam, and strong coffee. This would be our last square meal for twenty-four hours if all went well

Highline rigged across the Entrance Room to facilitate bringing in supplies. *Tom Culverwell*

inside the cave. But if someone got hurt, there was no knowing when we might get a hot meal again. It could take days for us to haul and hoist an injured companion out of this dark, dank hole in the ground.

We were all dressed in heavy, nail-studded mountain boots, mud-caked coveralls with well-worn leather insets where mountain ropes rub, and dented, scratched miner's hats. None of us were novices; this was no cave for beginners. Breakfast over, we tested our karabiners and checked every foot of the nylon climbing lines for rot or wear. Finally, we helped each other cinch on our packs. One man picked up a large coil of rope, hung it around his neck and settled it over his shoulder. Others did the same, and the party was off to the cave entrance.

We filed under the forty-foot arch of the cave entrance and stopped just inside at a wooden trough to enjoy ice-cold drinks of water. Our carbide lamps, the principal source of light on the expedition, were fueled. Our supplies were hauled across the highline, installed to avoid carrying them in our climb out of the Entrance Room. We were soon out of the twilight zone, and pushed on to inky darkness, where light is only a tiny pinpoint in the vast, black void. After negotiating a long, sinuous upper passage cut through cave clay, we arrived at the Jumping Off Place, the entrance to the Big Room. This was the brink of the first chasm, where light from our headlamps seemed to fade to a gloomy nothingness. The far wall of wet dripstone was well beyond even the beam of the most powerful flashlight we possessed.

As we climbed down over the first, twelve-foot pitch, Hyman, our strong, wiry leader, stopped us with terse instruction to tie-in. We each fastened a short line to a rope tied about our waists and then secured the other end to a projection of the wall. In this way we made sure that if someone were jostled off the ledge, he would not fall very far.

John, our safety man, found a spot behind a piece of projecting rock where he could hold the first climber on a safety line, should the climber slip. When he tested the line, he perceived a slight movement of the rock. Sharp inspection proved that the rock was only a large boulder embedded in the mud on the ledge. Realizing that the boulder might be pried from position and go crashing down on a climber below, John took off his carbide lamp to smudge in large carbon letters the words "Loose Rock" on both sides of the boulder. He then fastened his waist rope to a piton driven directly into a crack in the rear of the ledge.

Joyce, Carl, and the rest of us clung and huddled on this narrow shelf scarcely big enough for six. Hy inspected each knot before roping himself into the depth below. Satisfied that everything was safe, he drew the climbing line around his body, in the standard rappel position. He stepped backward off the cliff and slid down the line. His job was to seek out the safest way over the Grotto down to the floor of the Big Room. We watched his light dwindle farther and farther into the gloom, as John's hands steadily played out the lifeline snubbed around his body.

Hy's voice echoed up to us from below in meticulously separated syllables, "I-am-nine-ty-feet-down-at-the-end-of-my-rope. Send-down-an-other-line."

John stood steady and alert, in case he had to hold Hy on the safety line, while I swung the loops of the second line out over the edge and down into the blackness. Hy caught the slithering end,

SCHOOLHOUSE CAVE
PENDLETON COUNTY – WEST VIRGINIA.

THE SMALL CIRCLES INDICATE STATIONS ON THE SURVEY AND ARE IDENTIFIED BY NUMBER IN THE ACCOMPANYING ARTICLE. IN THE SCALE OF THE MAP THESE CIRCLES ARE SIX FEET IN DIAMETER. BEING APPROXIMATELY THE HEIGHT OF A MAN, THEY WILL GIVE SOME IDEA OF THE PROPORTIONS OF SCHOOLHOUSE CAVE.

SCALE IN FEET

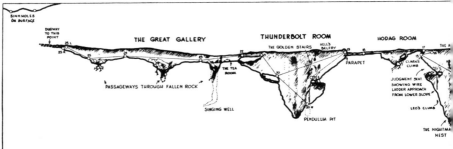
ELEVATION THROUGH STATIONS 8-25

Assault on Schoolhouse

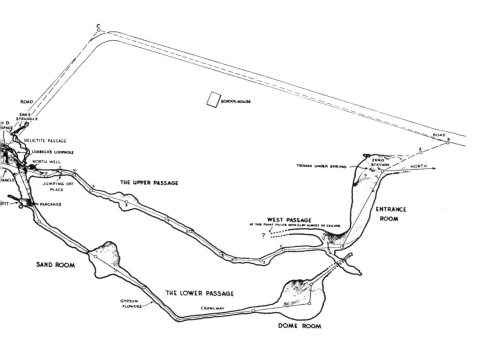

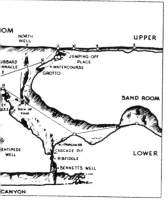

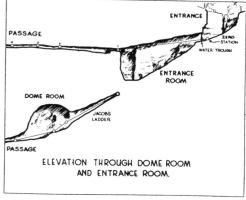

ELEVATION THROUGH DOME ROOM AND ENTRANCE ROOM.

carefully tied the two lines together, then gently tugged on the line. I let go of the other end. There was a whisper of falling rope, and then Hy was free to go on. Again the strands of nylon were sliding through John's gloved hands as Hy went down. Finally we heard a voice faintly echoing up the rock walls: "Down. Off-rope." We knew that at last Hy's feet had hit on the solid floor.

"Send-Ida-down-next."

I adjusted my helmet, checked my carbide to a faint half-inch of clear, yellow light, inspected a flashlight thonged to my wrist, and stepped into the rappel.

The safety man softly said, "On belay, climb."

I replied, "Climbing," and stepped backwards off the ledge.

I played out my line slowly with my left hand, which was behind my back. My right hand grasped the rope in front of me. My boots were flat against the vertical wall, and I was leaning out from it far enough so that I seemed to sit in the loop made by the line as it passed around my body. My nail-studded boots were playing a rhythmic tattoo along the vertical wall as I descended. The light from my lamp was little comfort in the inky vastness, but the flame had to be kept low. A high flame could easily burn the rope on the rappel or the safety line, and cause me to fall to the rock-strewn floor two hundred feet below. The carbide lamp is a convenient, inexpensive source of light underground, but it is an open flame. Not only the rope, but skin and especially women's hair must be kept away from the lamp.

The voices from above got fainter as I went down. The headlamps faded to tiny pinpoints of light. I sat confidently in the loop of rope, beginning to enjoy the feel of a new, strange cavern. The darkness closed in like a friendly tent pitched in a familiar place. I continued down until, at ninety feet, a different sound under my boots warned me that there was an overhang.

Suddenly I found a projecting wall to kick against. The free rappel is fun and easy when one has the confidence bred of long practice in simpler surroundings. However, getting down over the edge of the great overhanging shelf of limestone required a bit of careful maneuvering. Gone was the pleasure of being alone in the velvet dark, and I was glad to know that John was above, handling the safety line, and that Hy was below and had already negotiated this tough bit of rockwork before me. The safety line was evidence that the belay was still fastened above. Once over this great canopy of rock, there is a ninety-foot free drop through space to the floor of the Big Room. Now I realized that, although I would be dangling all alone in the middle of nothing, I was part of a well-trained team, taking my turn at enjoying the thrill of a dangerous adventure.

No sound then, but the slight, spluttering hiss of the little lamp and the slither of the rope through the hand that controlled the rate of my descent. Even my regular breathing seemed surprisingly loud in the black void. Gradually I became aware of heat as the friction of the rope warmed up the leather patches over my buttocks and back. When the heat became uncomfortable, I halted my descent and instantly felt John snub me on the safety line. I brought my flashlight into play as I hung in my sling. The light spilled out in each direction, but was swallowed in the gloom.

I let out on the rappel line again, and began to descend, but more slowly now, for the leather patches were hot. In a few more seconds, the nails of my boots struck the boulder-strewn, mud-covered floor of the Big Room. First I saw a glow, then a lamp on a helmet, and finally, Hy's figure became visible in the dark.

I called out, "Down," then untied the rope from my waist and slowly enunciated, "Up-rope." I then turned up my lamp and waited to assist the next climber down, thus releasing the leader to reconnoiter the Big Room while the rest of the party descended.

As the spelunkers came down, one by one, the room gradually brightened with their lights, and they, too, saw the towering walls around them, the sloping mud floor, and the climb that was ahead of them. It would be all up and down traveling here. Although Tom Culverwell's map showed an almost level ceiling for a half-mile (as the bat flies) to the back of the Great Gallery, we were not equipped with bat wings and so must pick our way across the Thunder Pit, over the Gargoyle Well, climbing back up to that impossible ledge, the Angel's Roost, then right down again to swing across the Pendulum Pit, before we could reach the back of the cave.

Carl bounced off the rope, rubbed his hands together, and said, "What a hole!" He found trouble at once in the form of a fissure leading down behind a huge boulder. He dropped himself halfway into it. Hy saw him and yelled sharply, "Don't go down there without a safety; how do you know how deep it is?"

"Aw, I can see bottom from here."

"Wait until I go first," he was reminded by the leader.

Hy, tied into a safety line, handed the other end to Carl and disappeared down the hole into the darkness. At about twelve feet, he called out, "Standing on a ledge." Suddenly the rope began slipping through Carl's fingers. Gradually he tightened his grip and leaned back on the rope.

"Are you all right?" he inquired.

"Lower me another thirty feet," said Hy. "The bottom *you* saw just crumbled under my boots, and I dropped four yards. I can chimney back out, if you let me down to the real bottom of the pit."

Now Carl carefully tended rope, keeping pace with the progress of the leader as Hy pressed shoulders and knees against opposite sides of the vertical passage. Hy finally appeared at the top of the hole.

"Want to go down, Carl?"

"No, you've seen it. I'll stay here until we go on to the Hodag Room," Carl answered weakly. (Spelunkers know that a hodag is an animal that has legs shorter on one side than on the other, so that it can negotiate its hilly native terrain without walking lopsided.)

By now, eleven of the cavers had arrived in the Big Room and were wandering around or directing lamp rays futilely upward into the complete darkness. One chap was busily setting up a camera on a collapsible tripod. One of the girls took a bite off a small hunk of cheese and poured water from a canteen to wash it down.

No one moved out of range of the voice of the leader, for underground, as on board ship, the decision of the captain is absolute.

Hyman was talking on the newly rigged telephone to two of the party we had left on the ledge. They had piled blankets, coffee, and first-aid gear back at the Jumping Off Place, and they reported they were in communication with the surface.

We left the rappel line hanging from the Grotto above; this was to be our last tangible contact with the outside world for some twenty-four hours. We had the field telephone, which could give us communication with the surface, but with its unwieldy reel, it was excess baggage. One cannot climb out of a cave on telephone wire.

It is a modern convenience that the early explorers of Schoolhouse did not use. Although we felt less isolated, thanks to the phone, we could not escape the fleeting idea that this link with the surface detracted a little from the adventure we sought here in the first place.

Rappelling down the Nick of Time into the Big Room. *Tom Culverwell*

Hy was all business as he led us, swinging past a vertical ledge, rightly named the Nick of Time, through a tight place called the Keyhole. We next swung across the Big Bite, a huge piece of rock which sloughed off the ceiling and was wedged between the two walls. It forms a natural bridge for the light of heart and the strong of stomach, for it is suspended high in the air between the Gargoyle Pit, forty feet deep, and the Thunder Pit, seventy feet deep. One by one we stepped gingerly across the great rock. One by one, we noted the sloping sides and made sure they straddled the ridge. John had a hard time on the Big Bite because he carried an extra line at his waist. It was trailing out behind him and seemed especially designed to foul on every projection. But this line had to go through, for it was destined to be the highline stretched between the two ends

Out of the Gargoyle Pit and over the Guillotine Stone toward the Judgment Seat. *Tom Culverwell*

of the Big Room. Along this highline, we would slide our gear, urging it along by snub lines at each end. A pack can seem as heavy as Sinbad's old man when one is traversing a delicate climb. So we were glad to be able to rig such a line to slide our equipment from one high point to the next, well above the floor of the cave. Although no one mentioned it, this, we all knew, was the procedure we would have to use if one of us were to be injured in an accident.

Hy got the first half of our party across the pits, and then he tackled another chockstone. This one, called the Guillotine, is wedged in a vertical position with a knife-edged ridge at the top. He worked his way up until he had his two hands over the knife-edge. Then he swung his body away from the rock; with a mighty heave, he raised a leg and landed astride the ridge. He worked his way cautiously along the ridge and then signaled for me to be the next climber. I found a good foothold, and swung up and out. My hand flew up for the knife-edge, but I clawed the rock an inch short of the handhold. Instantly, my safety line tightened, holding me in place while I grasped with the other hand. I found a new grip, and squirmed up to another foothold. As my leg straightened out, my arms, then my head came even with the knife-edge. This was enough. I swept one leg way up over my head in a movement I learned when I studied ballet as a youngster. When my heel hooked over the knife-edge, I hung suspended by that one heel and two arms, but only for a minute. A sudden arch of my back brought me up to the ridge.

One at a time, the other spelunkers climbed up over the Guillotine. When the last one of us was over the chockstone, Hy moved back to begin the assault on the Inner Wells. Working up a vertical slot, he pressed his feet against one side of the crack and pulled against the other side. The mountaineering manuals call this sort of climbing "laybacking," but no book could ever tell of the extreme pressures, the muscular fatigue, and the cramps that result from this method of climbing up a straight wall.

About forty feet up, Hy ran out of crack. It went no higher. I played my light up and down the wall. "There is a slight bump to your left about three to four feet," I shouted. Hy jammed his right boot in a crack and did the same with his right hand. Then he made a fist of his hand and wedged it firmly in place. Swinging his body out to the left, he explored the wall blindly with his left hand and foot. Finally, when he was stretched out over five feet, he breathed a sigh of relief that was easily audible forty feet below. He had found a little ledge. Quickly he stepped over to it and hung, balanced on one foot.

His hands searched above his head for another hold, but none was to be found. My light picked out a crack six feet above Hyman's head. He pulled out a rock drill and his piton hammer and went to work. At once the rock chips began to fly around him, bouncing off his safety glasses. Shards of stone rained down in a cascade on my upturned face. I brought my glasses down by a shake of my head, for I dared not let go of the safety line to Hy, or even take my eyes off him.

After ten mintues of hard work, Hy succeeded in placing an expansion bolt in the wall, and to this he fastened his safety line two

Traversing the Inner Wells from the Groan Box to Angel's Roost. *Tom Culverwell*

feet above his head. After the line was clipped to the bolt by an oval snap-ring of steel called a karabiner, Hy was ready to climb.

With a mighty bound from the ledge, plus a hand grip on the expansion ring itself, he pulled himself about three feet further up the wall. I pulled the safety line tight under his armpits, grunting as I took the weight of the 185-pound climber on the line. I leaned back into the loop of the line and placed my boots flat against the wall. Hy was held well up on the wall as though he were pinned there by some giant hand.

Hyman reached for the crack over his head and easily pulled himself up to it. As he rested, I started the climb. John had moved up to the base of the wall and held me on the rope as I came to the expansion bolt.

After more delicate, hazardous climbing across the Inner Wells and the Groan Box, we found ourselves at the start of one of the most famous traverses in underground America. We were a hundred feet above the floor of the Big Room, looking out on forbidding

Angel's Roost. Hy began to move slowly across this foot-wide ledge, while I carefully paid out the safety rope—the slightest jerk could send him hurtling down in a giant pendulum arc. After what seemed hours, Hy, then I, reached the safety of a high, narrow balcony, the Judgment Seat. When a cable ladder had been rigged and the others had negotiated the precarious traverse, Hy climbed to the entrance of the Hodag Room. Here he said, "John, I'll take some of the group on through the Hodag Room. You haul in supplies from the Jumping Off Place and then bring on the rest of the party. You are in charge."

Fifty feet above the Judgment Seat and two expansion bolts later, in the Hodag Room, the long arms and strong fingers of Hy came into play. Here, we traversed thirty feet across a crack between the ceiling and the sidewall. A slight shelf had been built out by lime deposits at this point, and Hy swung across it. There were no footholds, and the wall was too steep to permit using the body for friction, so he clung by his fingers. When a guide line was lashed in place, the rest of us crossed one by one. We felt like monkeys,

Looking down from just below the Judgment Seat. *Tom Culverwell*

Traverse across the ledge in the Hodag Room. *Tom Culverwell*

swinging our feet and clinging to the narrow rock moulding by our finger tips.

Precarious as this traverse seemed, it was probably safer than the original route through the Hodag Room. This was a delicate balance climb on a lower ledge, followed by climbing holds on a few old, brittle stalagmites on the slope beyond. Each member of the party carefully traveled the Hodag Ledge and then silently watched his teammates make the tour without mishap.

We were not just climbers now, but a team, matched in skill, each involved in the thoughts and activities of the others. Because we thought alike, talk was reduced to a minimum. Along the line of climb, there was no idle chatter, only the necessary directions and commands and a final, "Well done," from Hy.

Together once more, we started to lope along the easy, low passage to the next room. This level passage, after seventy-five feet, came out near the ceiling of a room 150 feet long. This was the Thunderbolt Room.

Kicking off to swing to reach the opposite ledge while rappelling over the Pendulum Pit. *Tom Culverwell*

We descended eighty-five feet over a dirt slope to reach the very edge of the Avalanche Pit. Rappelling, one by one, we kicked away from the wall, swung across the pit, and landed on a tiny ledge in a notch beyond. Hy pushed out with one mighty bound. I kicked too gently and had to swing back for a second try. Joyce took off with a gay whoop, and swung back and forth three times for the fun of it. The Pendulum Pit was just beyond, and here, too, we had to swing across the depths on the end of a rope. High above us towered Hell's Belfry, and above that, two hundred feet from the floor, was the Golden Stairs.

We continued to clamber upward toward a projecting rock called the First Balcony. After regrouping, we continued climbing to the Second Balcony, where we finally glimpsed the last tremendous room, the Great Gallery. We still had to cross the Tea Room, a fairly easy one. Then we negotiated another pit, this time the Singing

Well, to get to the Great Gallery traverse. It was perhaps 350 feet across this room.

Part of the time we were seemingly on the cave floor, but suddenly we skirted pits again on a series of shelves of thin rock, vertical partitions, and arches. I thought, "This is just like walking across a huge lace doily."

As Hy blazed a trail across the room, we followed, moving one at a time, slowly and cautiously, until we were all on the far side. We climbed up and up, until we saw a jam of huge boulders reaching to the ceiling. These rocks probably rolled in from the surface sometime in the distant past. Here we were practically at the ceiling and just a few feet (actually seventy) from the surface and the hot July sun. We had reached our objective. The back passage of Schoolhouse had been conquered.

Our elation was tempered by fatigue and the sure knowledge that all we had traversed must be done again to return to the surface, and this time we would have to fight our way up 350 feet since we were now that far below the mountainside entrance. We perched in a circle, resting before the return trip.

I untied one pack and dug out several tins of army rations. I sliced off a piece of Spam and passed the tin to the next person. Each of us was aware that there must be enough for eleven people, since John was just arriving, leading his party up the last slope. Each caver took only one bite. All except one, that is, who was too worn-out to remember that he was still part of a team. He grabbed the half-tin of meat and attempted to wolf it all down. A hand reached over his and firmly stopped him, while a quiet voice reminded him that there were others yet unfed. The next tin of meat was cut into equal parts before it was passed around. This man would be carefully watched by his teammates, who knew that he might be near his breaking point.

In the face of extreme fatigue, iron self-discipline is required here in the nether world. Not physical stamina alone, but perseverance to keep putting one foot in front of the other when one more step seems impossible, or worse, not worth the effort. Here is an environment completely strange. There is no sunlight, no moonlight, no wind. There are only hard surfaces to touch, only muted colors to look at. Each man stands on his merits alone, and he must be able to carry his part, and a little more, if he is to command the respect of his fellows.

Even the sounds are different in the heart of the rock. There are no chirps of birds nor sounds of distant traffic—only the scrape of boot nails along the hard rock, the constant drip of water seeping down from above, the sharp crack of a hard hat hitting an overhanging rock, the hiss of a flaring lamp, the brush of denim on the walls of the narrow passages. Voices, too, reflect this environment. One is more likely to speak in hushed tones than to shout. Perhaps the hushed tone indicates a sense of achievement, or a feeling of insignificance in the shadow of Nature's superb creation. There is another reason for keeping the voice low—a loud tone might dislodge loose rock overhead.

Reluctantly, we finally picked up our gear to leave the Great Gallery. The trail was more familiar now, but we exercised even more

The Great Gallery traverse. *Tom Culverwell*

care than on the way in, for we knew of the traps which fatigue can lay for the unwary. Ropes that weighed five pounds were now twice as heavy—partly because they had picked up moisture and partly because our tired shoulders sagged more.

We picked our way across the delicate brackets along the sides of the Gallery, through the Thunderbolt Room, climbed and swung back across Pendulum and Avalanche Pits, less frightening now, since they were dangers past. Even the Big Room seemed less formidable. We negotiated the Guillotine, the Big Bite, and finally arrived at the Gallery facing the Nick of Time. Here we huddled, while half the team climbed on up the steep incline where we had rappelled down. We set up the highline and slid the packs across, singly this time, for we were too tired to pull a heavy load up to the Grotto.

One man was stationed precariously midway between the Nick of Time and the Grotto, so the packs would not get snagged on a protruding point of rock. This man could not be held well on a safety line from above, for should he slip, he would swing like a pendulum into the North Well to his left; therefore, he was tied in by a short length of rope to the rock on which he balanced. He kept busy and warm from exertion, but those waiting, cramped in the Little Gallery, were cold, and impatient at every delay in getting out the gear. The excitement gone, we were just wet, tired cavers, with no ambition except to get a warm meal and "hit the sack" after almost twenty hours of practically continuous caving.

At long last, the final pack went up, and the safety line was tossed over for our ascent. Once, it swished on beyond us, and we had to reach out across empty space to retrieve it. Carl climbed up, then turned to toss the carefully coiled rope back to the gallery. It missed, and this time we were dismayed to see it plummet into the dark well at the left.

But astonishingly, before Carl could retrieve it, Joyce grabbed the now slack highline, wrapped it around her, and swung out over the well in a great arc.

Hy yelled "No," but she was already out on the line. Fortunately, it was still secured above, and before the rest of us had time to realize what might have happened through this unplanned maneuver, she had seized the flailing safety line and swung back to the security of the gallery. The leader above swore softly, but he said nothing further.

This is the pinnacle from which a daring but inexperienced young boy lost his life on his first visit to Schoolhouse, in 1950. From the Grotto, he dropped a rope and climbed down, not knowing that his hundred-foot line was too short to reach the floor below. When he discovered he was at the end of his rope, he tried to climb back up, hand over hand. Just as he reached the top, the rope broke. He tumbled backward into the abyss and was smashed to death. Unfortunately his rope was only a sash-cord, easily cut on the jagged rock edges.

It was the job of the last two people to haul in the essential lines; these had to be left to get the party out and then retrieved as the last members left. John was next to last, and I was last. When we reached the Grotto, we saw Carl's feet and ankles above us. He was waiting to take the ropes as we slowly coiled them and passed them up. Everything seemed to be in slow motion now; we were so tired we moved automatically—or not at all.

John, almost always meticulously careful about his gear, pulled out his canteen to drink and accidentally dropped the cap. He looked wearily down at the twisting, tumbling cap, turned the canteen over slowly in his hand, and let it drop and roll down the slope. Only then did he realize that he was still thirsty. He watched the canteen as it bounced echoingly down to join the heap of flotsam in the Cascade Pit below—a sea floating a single overshoe, a battered mess kit, bits of assorted gear, all mute evidence of other hardy cavers pushed almost beyond endurance.

John and I slowly coiled the last piece of line and handed it up to Carl. John tied on the safety line and staggered up the last steep pitch. The line softly slithered down over the rock again. I tied on and was ready to climb out of the cave. But just before I started up, I paused a moment to savor the velvet darkness, the immensity of the cavern.

The Impossible Pit

The previous chapter was a true account of the successful though arduous exploration of a technically difficult cave. Perhaps all of Schoolhouse Cave has not yet been discovered, but its major climbing hazards have been overcome by many explorers.

James Cave in Kentucky presents completely new and, so far, insurmountable conditions. No one knows how extensive the cave is; its complex and tortuous passages leave invaders bewildered as to which section to explore next; and, most important, almost every foot of the way is difficult and perhaps impossible. Roger W. Brucker tells of the attempts of one group to overcome only one of these many problems.

Mr. Brucker is a co-author of The Caves Beyond, *the story of the Floyd Collins Crystal Cave expedition undertaken by the National Speleological Society in 1954.*

ROGER W. BRUCKER

Imagine yourself on a remote Kentucky hill in the year 1880, standing in the shade of big maple trees. Three men on horseback slowly approach, their shabby clothes dusty from the primitive roads. You slip behind a tree trunk and watch them dismount. One man sits down on a rock ledge, slowly pulls off a worn leather boot, and waggles a bare foot in the sunshine. The other two men cast quick glances around them, then stealthily pick their way over fallen logs to a limestone ledge, where a low, oblong hole leads into the blackness of a cave.

Twenty minutes later, the two men return. They exchange words with the seated friend, then all mount their horses and ride off. You never see them again, but you hear a lot about them—Jesse and Frank James, two of the most notorious outlaws in the history of the American Midwest.

If legend were truth, you might have witnessed this very occurrence. It is not difficult to picture such an incident, for the James brothers knew the cave country of Missouri and Kentucky well.

As years went by, their supposed rendezvous became known as Thousand Rooms Cave to the few hardy cave explorers with coal-oil lanterns who "made the rounds" of Kentucky's caves in the 1920's and 1930's. But Thousand Rooms Cave was bypassed in that heyday of commercial cave development. It contained no formations of great beauty, only a few dried and spalling stalactites of a dirty brown color, and even these were difficult to reach. Onyx miners, ruthless vandals created by the souvenir trade, also bypassed Thousand Rooms Cave for other, more beautiful caves which could be stripped for a quick dollar. In the years after World War II, cavers came to know Thousand Rooms Cave as James Cave. We have no real evidence that the James brothers ever frequented James Cave, but folk-say claims they did so. In any event, we know a good deal about several other visits to it.

On an April day in 1952, the quiet woodland scene was disturbed by the roar of a jeep, a bright yellow jeep, with seven people clinging to it. By the time these explorers and their equipment were unloaded, a file of other people had joined them, and the whole group moved through the trees and over the limestone boulders to the entrance to James Cave.

These cave explorers were members of the National Speleological Society, meeting in nearby Louisville for their national convention. Almost everyone wore coveralls and hard hats. There were some whose shoulders sagged beneath large coils of rope; others carried bulging canvas packs containing flexible cable ladders.

The first file of explorers stooped under the ledge and walked down a steep mud incline into a high room, reminiscent of a gloomy Gothic cathedral. Deliberately they passed by a dozen openings as they made their way to a place where the main passage turned left. The leader of the party directed them into a low stoopway, which rapidly pinched down to a crawlway barely eighteen inches high, with a dusty gravel floor. The eastern cavers were used to such inconvenience since their caves are small; they joked about driving all this distance to see a "New Jersey" cave. But after crawling the length of a football field they reached a sudden drop-off over a hundred feet deep. The only way across it is along a two-inch ledge.

Below yawns the Judgment Pit. There were fewer jokes then. Several decided not to attempt the crossing. The experienced rock climbers in the party found finger and toe holds and quickly crossed to the far side.

A quarter of a mile from the entrance, the diminished party duck-waddled beneath a low ceiling along a floor strewn with small, flat slabs of rock. Then the leader stopped them; he had come to another pit. No ledge borders this pit, and lights failed to reveal its depth. He hurled a slab into the darkness. Seconds later, the silent group heard it strike bottom and shatter into fragments.

Across the pit and around a cornice to the left, the passage continued. A safety line was tied around one man who volunteered to reconnoiter a route to the far side. With three men belaying the rope, he moved carefully down, over the loose scree to the precipice. He probed the blackness with the beam of his flashlight, then played it over occasional projections of rock.

"It's impossible to get across."

"How about with pitons?" called one explorer.

"No cracks to drive them into," said the man on the rope.

An experienced rock climber scanned the wall and agreed, "It's impossible."

"If they can't cross the pit, perhaps they can reach the bottom."

"No," said the leader, "there isn't time. Besides, one explorer from Louisville has already gone down 120 feet into it and we don't have that much rope."

Slowly the party started the return trip.

On Thanksgiving Day, 1953, another group moved into James Cave. This was a reconnaissance party crawling in to assess the possibility of a later crossing of the pit. The leader of this party of three was one of the explorers turned back by the "impossible" pit. One man carried a twenty-five-foot length of rope; another a camera and flash bulbs; the third, a gasoline lantern. At the brink of the hole the lantern was turned on. Its brilliant white light illuminated walls which before had been seen only in the feebler glow of yellow carbide lights. Perhaps with this added light the explorers would be able to find hand and toe holds hitherto overlooked.

There are none; the walls are as smooth as a concrete elevator shaft. Thirty feet across the pit the passage continues. The photographer "shot" the pit from every angle. Each man in turn descended to the "Diving Board," a pinnacle of bedrock jutting into the shaft, but eventually each was forced to return. The only way to cross, they decided, would be on a bridge, but who ever heard of using a bridge to cross a pit in such a cave? Anyway, how could a bridge be brought in through the twisting crawlway? Nothing over four feet long could pass the tortuous bends, and the bridge would have to span at least thirty feet. It still seemed impossible.

For months the three men puzzled over the problem before they decided how a crossing might be made. Nearly a year passed before they were ready to tackle the pit.

Finally, on Labor Day weekend, 1954, three men arrived in a truck with a ton of special equipment and food. Under the tarpaulin

One of the explorers rappelling into the pit on nylon line. *William T. Austin*

on the truck were nearly four hundred feet of stout rope, cable ladders, a drum of carbide, pots and pans, many odd-looking sections of water pipe, refrigerator boxes, and a score of prime-grade steaks.

One of the three men was Roger McClure, Ohio State University student and cave surveyor, lean, muscular, and red-haired. The second man, Tony VanDerLeeden, from The Hague, Netherlands, was in the United States Air Force, as was his friend, Robert Palmer. These three men made up the advance party of an expedition of the Central Ohio chapter of the National Speleological Society. The rest of the group was scheduled to arrive after midnight. In the meantime, the survey party was to plot a baseline from the cave entrance to the "impossible" pit, then draw a map from which the leaders could assess the probability of finding an alternate route to its far side.

By eight o'clock that evening, the three men had established one surveying point or station after another to the very brink of the pit. Back outside, McClure plotted a map on a clean sheet of tracing paper, while the other two slept. Finally he placed his dividers and scale in a case and crept into his sleeping bag.

At three o'clock in the morning, several station wagons arrived, bringing thirteen cave explorers who had been driving all night. McClure climbed out to welcome them.

"Do you have the map?" asked Philip Smith, tall, sturdily built leader of the expedition. He studied it over, biting his tongue in deep concentration. Then everyone settled down to get some sleep in preparation for the day ahead.

The smell of bacon and eggs revived the party. Photographer Jim Dyer's cameras were already clicking, catching the explorers off guard: Bill Hulstrunk, a young chemist who flew out from New Jersey to take part in the expedition; Fred Mysz, a Chicago lithographer, with a blue-black "5 o'clock shadow," pulling his red hard hat out of a duffle bag; the cheerful bustle of an early working camp.

Behind this scene lay three months of planning, a half-dozen reports, a frantic period of equipment-building, and a preliminary field test. Hundreds of dollars had been spent to put the group where it was, in position to strike at the "impossible" pit. Burnell Ehman, a machinist from Yellow Springs, Ohio, had designed and built a bridge of water pipe to stretch across the black gulf. Composed of four-foot sections—to be carried in separately and to be assembled at the edge of the pit—it had been designed to support three times the weight of a man.

A month earlier, Smith had called the group together for a field test of the unusual structure. Along a gorge whose sheer walls dropped forty feet to the water the explorers pushed the insecure-looking pipe bridge toward the opposite cliff. When it was securely fastened with guy ropes, Smith clicked the clip on the karabiner on his safety belt onto a welded steel ring designed to slide along the pipe as he moved.

Hanging under the pipe like a sloth, he slowly worked his way across to the other bank. Those were tense minutes, but the bridge passed its test with admirable steadiness. Now all was ready for James Cave.

Crossing the pit was the first goal of the expedition, but there were other objectives; one was to map as much of the cave as possible within the two-day weekend; another was to study all aspects of staging a medium-sized cave exploring expedition, a subject on which little research had been done. Whatever methods and equipment proved serviceable here would probably become standard for further serious explorations.

Now, in November, a strange procession, laden with sections of pipe, rope ladders, telephones, wire, tools, and food, moved toward the entrance of James Cave. On this occasion, I was a member of the expedition. With lighted lamps, the explorers pushed into the cave and stashed their equipment at the mouth of the eighteen-inch-high crawlway. Working as a team, they relayed the supplies forward, unmindful of the bats that swept back and forth above them. "Clank, clank, clank"—section after section of pipe passed from sweating man to sweating man, each lying prone in the tight-fitting limestone tube. By noon, all equipment was assembled in the low, flat-ceilinged passage near the edge of the pit.

With sledge hammers the explorers drove rock drills into the tough limestone overlooking the pit. When one man tired, another took his place, and the steady rapping of the hammer reverberated through the passage. After two monotonous hours, four eyebolts were twisted home in their lead anchors. Then one team began to bolt together sections of the bridge, while another untangled the block and tackle.

Smith and I moved out to the rock promontory as the pipe clanked forward. Both of us were "tied in" to eyebolts with stout ropes, but nevertheless we were careful to keep as firm a footing as we could while we tried to guide the free end of the pipe toward the mouth of the virgin passage on the far side. Rocks, dislodged by our feet,

Expedition leader Philip Smith, fastened for safety, drives rock drills into the tough limestone. Lanyard attached to sledge hammer prevents losing it in the depths of the "impossible" pit immediately beyond. *James Dyer*

pelted down the shaft, ricocheted off the walls, and at last hit the bottom far below.

After three hours of alternately pushing the bridge out, pulling it back, adding a section, and pushing it out a little farther, the group turned to their leader. They were exhausted, and so far had little to show for their efforts.

There was still tomorrow—but we were eight hours behind the schedule we had hoped to make.

After a supper of steak and baked potatoes, the leader held a council of war. "Tomorrow," said Phil Smith, "we'll locate the position of the pit on the surface. Perhaps we can find a back door, another entrance. While Bill is doing that, Fred and I will rig the pit for descent." Dusty, tired, and discouraged, the cave explorers crawled into sleeping bags.

Next morning, our survey party sighted along the surface, past rhododendrons and maples and sinkholes, and dropped a rock on the ground to indicate that the pit was directly below. The rock lay at a place where the hill dips toward the valley—the edge of the escarpment.

David Jones, wire chief, Burnell Ehman, and I found another small cave nearby, but the end was plugged with mud. That didn't help. After surveying the small cave and exploring a side passage on the way to the pit, we joined the other party.

Smith and Hulstrunk were at the bottom, yelling that they were famished. Micky stuffed cans of food into my sack. I tied a not-too-snug bowline around my chest and checked with my safety man, Fred Mysz, who sat beside a two-hundred-foot coil of rope. He was tied into an eyebolt and backed up by David Spitler, a husky Ohio football player. I let Fred know the signals I would be calling back to him, and we double checked them.

The ladder, anchored to an eyebolt, stretched over the rock slope,

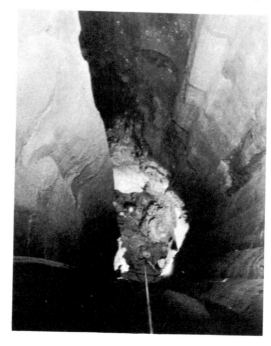

Even from a point midway down its shaft, the "impossible" pit challenges the most experienced explorer. *William T. Austin*

then plunged into the pitch. My heart was pounding, for I had never used a flexible cable ladder before. From the brink, I could see a small pool of yellow light where two tiny figures stared up at me, and I warned them to take cover for protection from any rocks I might dislodge. The descent seemed endless, but finally I stepped onto the bottom. Phil Smith announced that we were 130 feet below the others. The ceiling of the dome was about twenty feet above them, so actually the pit is about 150 feet deep. We looked up at the feeble glow of the gasoline lantern eleven or twelve stories above us. Behind us, a muddy tube turned a corner on its way down.

"Where does it go?" I asked.

"No place," said Phil.

After we had warmed canned hamburgers over our carbide flames and munched candy bars and peanuts, Phil moved over a pile of muddy rock to a place where a parallel pit plunged even deeper than the pit we were in. We climbed down thirty-five more ladder rungs to the bottom, onto a circular gravel floor, and poked our heads into a narrow crevice going down still farther. The clatter of a test rock indicated that it must be at least another fifty feet to the bottom of that—perhaps more.

But we had run out of ladder. This additional fifty feet or so makes the "impossible" pit approximately 240 feet deep, the deepest pit we know about in Kentucky. We placed our initials unobtrusively on the wall. This was not a defacement of natural beauty, but merely a sign of our having reached this far, forbidding spot, which few will ever see.

Climbing out was a painfully slow process that drained all our energy. We spent anxious moments rekindling carbide lamps that brushed out against the cables. I mounted one fifty-foot section in darkness because there was no secure place to stop to relight. Bill Hulstrunk's lamp went out five times, and five times he relit it. At the top, he stretched out on the jagged-rock floor, too exhausted to move.

The explorers assembled for one last attempt to push the bridge across. This was not a "do-or-die" attempt, for cave exploring is not a "do-or-die" sport. Exploration proceeds in spite of risks, not because of them, and then only after standard safety precautions are in effect.

Jim Dyer, with his camera around his neck, scrambled down the

A tense moment in the first attempt to bridge the pit. *James Dyer*

ladder to a place thirty feet below the bridge. The sagging steel pipe swung out over the void and was rammed into position on the far side by brute force.

Everyone knew that the credit for any success rightfully belonged to all the men who toiled and sweated for this moment, so one man was chosen to attempt the crossing, but all agreed to keep his identity a secret. That man, hooked into the steel ring, swung his body under the pipe, slothlike, and started his slow traverse. He was halfway across when suddenly his feet slipped off the pipe! A metallic clang rang out as his safety belt abruptly broke his fall. At that second a brilliant flash lit the pit; Dyer's camera caught the man in the terrifying moment when he dangled beneath the pipe.

After a few difficult seconds, the man managed to swing his feet back over the pipe and continue his slow journey. At last he reached the other side, unhooked himself, and disappeared into the unknown.

Before very long he was back. He had discovered another deep pit around the corner and to the right. Should he go on? Smith bit

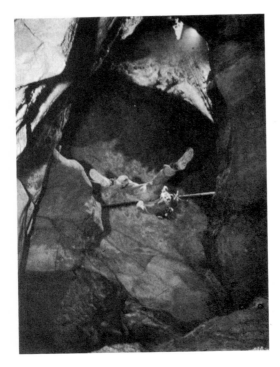

Unposed photograph as explorer slips on the pipe traverse crossing the pit. Dyer took the picture from a ladder dangling thirty feet below the pipe bridge. *James Dyer*

his tongue as he studied his watch. Outside, he knew, it was dark. He knew, too, that spirits and energy were at a low ebb, that all the equipment would have to be dragged out of the cave for the long journey home, and that tomorrow was another working day.

"You'd better come back," he decided. "Time is running out."

Around a roaring campfire that gave new strength to weary, aching bodies, the party tallied the score: McClure's survey party had mapped a little more than a mile of cave passages. In their probings into remote galleries they had found what appeared to be the main part of the cave! The Dutchman, quite by accident, dropped the party's lunch into a hundred-foot canyon and will probably never hear the last of it. To make up for it, however, he discovered a virgin grotto, adorned with flowstone and a massive stalagmite three feet in diameter and six feet high.

"It was yust a little place," he says, and "Dutchman's Inn" is the name of VanDerLeeden's discovery.

The other party had descended deeper than anyone else had ever gone in a Kentucky underground pit—a dubious distinction, perhaps. They had also crossed "Forty-Fathom Pit," the name the group coined for the "impossible" hole. A record? Yes, although it was rendered less meaningful by the fact that they had not been able to exploit their bridgehead. Most valuable of all was the experience gained in conducting a difficult expedition underground. The use of new equipment, better teamwork, and better planning would all pay off in future explorations. There were other compensations, like Jim Dyer's superb photographs of cave explorers at work. One special picture, Jim made while crawling along a passage not far from the brink of the pit. It showed two names scratched crudely on the wall in genuine old-style lettering: Jesse James, Frank James. Did this prove there was some truth in the old legend?

Since that day, another expedition has gone exploring in James Cave. The Kentucky-Indiana chapter went in with white nylon ropes instead of ladders and rappelled their way down. They found virgin cave, for the absence of footprints showed that no others had been before them. Not even the Ohio expedition had penetrated that passage. Down a series of boulders, over mud-covered rocks, slippery and treacherous, and they were in the bottom of a large canyon, perhaps forty or fifty feet high and about fifteen feet wide. They knew that they were where man had never been before. Then suddenly, a cul-de-sac, and the passage ended abruptly.

They had found the bottom of Forty-Fathom Pit. This was one area where cave explorers in search of new cave would not need to go again.

Or would they? Still another assault on James Cave is being planned. Have we really reached the bottom? And what about the main part of the cave that McClure's survey party discovered back in 1954? These are unanswered questions that nag the mind and challenge the spirit.

The Miners' Bathtub

The lure of unscaled mountain heights has spurred climbers to superhuman efforts, sometimes with fatal results. Since one man's weakness or mistakes can imperil an entire party, only the best-trained, the best team workers, are selected. If mountaineering demands special skills and unwavering discipline, so does difficult cave exploring.

Alpinists have the advantage of reconnoitering with telescope and aerial camera. Speleologists have no way of anticipating what new difficulties lie ahead, around each corner or over each ledge. Water greatly increases the problems—as is apparent in this story of the attempts to explore the Devil's Hole, in the Nevada desert.

Here, added to all the dangers of mountain climbing and cave exploring, is a new hazard—cave diving with aqualungs. Dr. Halliday's interest in cave exploration is so intense that—as a busy young physician, just out of the Navy—he found time to organize chapters of the National Speleological Society in Seattle, Denver, and Salt Lake City. Lieutenant Halliday was recalled to service late in 1954 and is now in the Pacific.

WILLIAM R. HALLIDAY

In the Great Basin between the massive Rocky Mountains and the sharper slopes of the Sierra Nevada of California, dozens of snow-capped ranges look down upon innumerable caves. They are quite different from the multitude of wave-cut shelters on the long-dry shores of ancient lakes on the other side of the mountains.

This is desert country, where caves of any kind seem out of place, and the large streams or deep pools, which the explorer encounters in many caves, seem utterly alien. Strangest of all, however, is the cave that contains a 93-degree hot spring. It is the Devil's Hole, seventy miles west of Las Vegas, in the midst of the Nevada desert.

The story of the exploration of Devil's Hole begins one bitterly cold night on the desert, late in January, 1950. That night, north of Las Vegas, Peter Neely and I complained through chattering teeth as frost accumulated on our sleeping bags: This country was too cold for camping.

Pete is one of the outstanding cavers of California. Not long before, he and I had been the first to reach the bottom of the fabulous Cave of the Winding Stair, not far to the southwest in the Mojave Desert of California. Now, our target was Gypsum Cave near Las Vegas, where ancient man was first found indubitably associated with the giant ground sloth, the American camel, and other long-extinct beasts like the sabre-tooth "tiger." Here we were to find that patches of the coarse, mossy hair and large droppings of the lumbering ground sloths, known to have been extinct for more than 12,000 years, were still visible on the floor of the cavern, preserved through the centuries by its amazingly dry air.

We knew that the quick trip we planned to Gypsum and the other small caves nearby would leave us considerable leisure for reconnaissance. Naturally enough, the bitter weather determined our route, and since Death Valley was the warmest spot we could think of, we decided to go there. Anyway, there were supposed to be a half-dozen caves in that area that had to be investigated eventually.

Next morning, we headed northwest from Las Vegas and Gypsum Cave toward Death Valley. While passing the snowy limestone summits of the Charleston Range, Pete and I took turns looking at the map. Suddenly, an odd name caught my eye—Devil's Hole.

We had seen the name on the map before, and had wondered about it. Could it be a cave? We knew that two or three other Devil's Holes in other states were also caves. The Devil's Sinkhole, Texas, is an immense, bat-infested pit, with a magnificent cave at the bottom of its frightening drop. On the other hand, another Devil's Hole, in southern California, is a small, dry flat, amid towering cliffs which raise the temperature almost beyond human tolerance in midsummer. This Nevada Devil's Hole, then, could be almost anything. We'd find out, we decided, for we'd never have a better chance. When we reached the approximate location of the dirt tracks that are Nevada State Highway 16, we chose a well-traveled pair of ruts and bounced happily off into the desert. After all, we were used to this kind of road, and had even built one to get to the Cave of the Winding Stair. After a few miles, our map showed a junction. We were making excellent time when the road suddenly reached a salt flat and broke up into a maze of tracks apparently made by a sheep wagon. For a while we seemed to be traveling in aimless circles. Then we saw the top of the wagon itself above the brush. We soon spotted the herder and hailed him.

"Devil's Hole? Sure. It's right through that pass over there," he assured us. "You can't miss it." We looked. A less passable mountain chain we had rarely seen.

"The Miners' Bathtub, they used to call it," he continued. "When they was lots of mines around here, all the miners used to go over there every Sattidy night and git cleaned up and shoot the breeze. Blindfish in there, too."

At this we really pricked up our ears. We are still looking for the first blindfish to be found west of the Continental Divide in the United States—but that is another story.

Despite our skepticism, we followed his directions and were pleasantly surprised to find the pass through the low, barren range without difficulty. Then, almost at the side of the car, a precipitous opening appeared in the level surface of the plain in the shadow of a small peak.

Excitedly, we piled out of the car. The pit ahead amazed us, even though we were used to caves and their remarkable entrances. The wall beneath us dropped almost perpendicularly for fifty feet, while the wall thirty or forty feet across the chasm sloped so steeply downward that not even the wild burros we had seen on the hillsides could have managed it. The surface opening was nearly a hundred feet long, and at one end, ages of surface wash had constructed a kind of stairway down the steep slope.

Below and to our right was an alcove perhaps thirty feet high and half as wide, separated from the surface by a twenty-foot ledge of solid limestone. The grotto tapered downward, and within its mouth lay a narrow pool about sixty-five feet long, sparkling as the slanting sunlight penetrated beneath the ledge. It shone crystal clear and aquamarine at the shallow end beneath the opening, but back against the rear wall, it gleamed with the incredible sapphire seen in cave pools of great depth.

We climbed down the "stairway" and examined the interior of

the grotto. It had the appearance of a limestone cavern carved out by the slow but inevitable action of subterranean water. Warmed by our exertions and out of reach of the chill desert wind, we sat down and relaxed in the sun.

As we watched the lazy ripples in the clear blue water, we suddenly noticed a school of tiny fish feeding at the edge of the pool, amid the debris washed down from the surface. Were these really the blindfish we had sought so long? No, their eyes were clearly visible, and they showed every indication of keen vision, darting off whenever we approached too close. Exquisitely formed, they looked like tiny blue-gray perch. How did these strange little fish happen to be here, far from any stream? At the time, we had no answer.

Along the eastern edge of the pool, the wall rose vertically to the ceiling high above. On the other side, however, the wall was less precipitous, and, a few feet above the water, there was a small ledge along which we could make our way. Behind the pool, the rear wall slanted down beneath the surface, but at its west edge, we found what we had hoped for—an opening extending back into complete darkness.

Unlimbering flashlights, we squeezed into the narrow, slanting chamber. Around us were the irregular patterns of solution, and a little flowstone and dripstone of a later period of deposition were apparent. But the passage quickly narrowed to a mere slit. We were blocked. There were tantalizing suggestions that more cave lay beyond, but no possible route existed above water, and we were neither equipped nor prepared for the difficult sport of cave diving, which has claimed lives in both Europe and the United States.

We took pictures, sketched the cave, and prepared to leave. Oddly, neither of us had bothered to test the temperature of the water. I was halfway along the wall of the cavern lake when an incredulous yelp came from behind me.

"The water's warm!" Pete yelled.

Spinning around, I charged back down the steep slope. Cave pools and streams are notoriously icy, and while a few thermal pools are known in natural limestone caverns in other countries, no others are known within the United States. There was no doubt about it though. The water of the pool was gloriously warm—almost 93 degrees Fahrenheit, we later learned. No wonder the miners came here every Saturday night. Yielding to the obvious temptation, we spent the next two hours floating luxuriously in their bathtub.

Actually, we did more than merely float. Spying an apparently new flashlight submerged on a ledge which seemed to be just below our toes, I dived for it, quite unsuccessfully. The crystal-clear water was even more deceptive than usual. Taking great breaths to fill my lungs, I tried again, kicking down and still further down, ears popping, to a depth later measured at twenty-five feet. With my last ounce of energy, it seemed, I grasped the light, kicked off the ledge, and triumphantly bore it upward as my lungs were about to burst. One look at it on the surface, and I threw the corroded thing away.

Nor was another enterprise more successful. As we studied the slanting rear wall and considered the small upper chamber, we became even more convinced that additional cave existed beyond the barrier. By diving deep enough, it should be possible to pass beneath the rear wall and find the hidden passages. We dived repeatedly to the limit of our endurance, but proved only that our special flashlights still worked when full of water.

It was with regret and frustration that we left, for we were certain that we would have achieved our goal, could we only have dived deep enough. Neither of us had any idea of the enormous depth which eventually had to be overcome before our prediction could be proved accurate.

Since their appearance and surroundings were so unusual, we collected four specimens of the tiny fish. A week later, Pete and I took them to the Scripps Institution of Oceanography, a division of the University of California. There, ichthyologist Carl L. Hubbs quickly identified them as a little-known desert spring fish. These "pupfish," he told us, are known as *Cyprinodon diabolis* in recognition of the fact that they are found nowhere in the world except in the Devil's Hole. Only twice had these tiny fish come to the attention of the scientific world. In 1930, the species was discovered and described by Joseph H. Wales. Later, Robert R. Miller, a biologist from the University of Michigan, had made a thorough study of the fish in this unique pool and those in many other small desert springs. Outside of ichthyological circles, though, the Hole was almost completely unknown.

Dr. Hubbs explained that these tiny fish had descended from a better known cyprinodont fish which was widely distributed in a chain of lakes which existed in the Mojave Desert during the moister climate of ancient Pleistocene times. Then geologic uplifts occurred, and the lakes in the part of the desert northeast of Death Valley were either drained by down-cutting of their outlets, or they evaporated as the climate became hotter and drier. The lake that occupied the valley containing Devil's Hole thus disappeared, but a few springs and the unique cave pool remained in spite of the retreating waters. In these refuges, a few of the pupfish survived.

In some parts of the Mojave Desert, the lakes did not vanish until as recently as 11,000 years ago, but isolation of the Devil's Hole probably occurred much earlier. During their thousands of years of isolation, a succession of minor variations produced major differences between the scattered groups of fish. Those in the Devil's Hole became so different from their cousins in other desert springs that they provide an excellent demonstration of evolutionary modification.

On Dr. Hubbs' advice, the word was passed among cavers that these fish were not blind, that they had already been scientifically studied, and that their hold on life was exceedingly tenuous. Since no reason for further collecting existed, it was voluntarily banned by unanimous agreement among cavers. In addition, Dr. Hubbs alerted the National Park Service to the need for protection for these unique little fish and their equally unique home.

Our report of Devil's Hole raised a lot of questions. How deep was it? If there were underwater extensions, how could they be explored? Unfortunately, I had to be absent from California for several years, but Pete and other cavers spent many hours discussing further investigations.

In June, 1950, a party from the Southern California Chapter of the National Speleological Society, led by Walter S. Chamberlin of Pasadena, "blazed the trail" into the unknown depths of the Hole. Wearing a diver's helmet, Walt reached a depth of seventy-five feet. Somewhat to the surprise of a few skeptics, he found no boiling springs as he descended. The main hazard to further penetration came from dislodgment of loose rocks by the diver's lines trailing along the steep slopes and ledges.

At this depth, only a faint reflected blue glimmer broke the darkness, but with the use of an improvised underwater light, Walt

The canyon entrance of Devil's Hole leading down to the Miners' Bathtub. One of the explorers is about to make the first descent into the unknown depths of the hole using a diver's helmet. Later descents were made with aqualungs. *Walter S. Chamberlin*

was able to make out the openings of several possible passages. Unfortunately, the clumsiness of the unwieldy diver's helmet precluded his following even the shallower leads.

Exploration was thus proved humanly possible and very desirable, but little progress was made for three years. One important discovery was made, however, a short distance northeast of Devil's Hole. Here was found a small, vertical, pit-like cave 130 feet deep, which terminated in a small pool. Its temperature was similar to that of the Hole, an indication that there was probably some kind of communication below the surface. This smaller cave was named Devil's Hole Cave.

During this time, the Devil's Hole had been made a detached part of Death Valley National Monument, and the National Park Service thereby provided protection to the site. That collecting was forbidden was fortunate, for the publicity resulting from the establishment of a national monument caused the existence of the Hole and its tiny inhabitants to become widely known. The Park Service, however, was almost as anxious as the cavers to learn the secrets

A protective gate placed in Devil's Hole entrance provides security for the rare cave fish, *Cyprinodon diabolis*, found no other place in the world. This detached portion of the National Monument is in Nevada. *Walter S. Chamberlin*

of the Hole, and permission for continued exploration was cordially extended to the Southern California Chapter.

In the spring of 1953, William Brown and Edward Simmons of Pasadena decided to use aqualungs to conquer the depths of the Hole. Simmons and Brown formed a remarkable team. While Brown did the actual diving, it was Simmons who purchased several aqualungs and who developed the special gear which made the future explorations possible.

For several weeks, members of the Southern California group practiced with Simmons' equipment in the swimming pools of long-suffering neighbors and at the beaches to the south. Such procedures as the emptying under water of a half-filled face mask, the re-insertion of a dislodged mouthpiece, and the adjustment of air controls were rehearsed until they became routine even at considerable depth. When these fundamentals were mastered, the cavers began to learn more advanced techniques, such as two swimmers using a single aqualung.

Still, as the scheduled day for the great attempt approached, innumerable questions remained unanswered. In only one other western cave—Bower Cave, California—had aqualung diving been attempted, and the newspaper accounts of Jon Lindbergh's dives beneath the relatively shallow barrier there explained very few of the hazards to be encountered in this brand-new technique of cave diving.

Nor could reports be found of any previous dives into warm or hot springs. Would the higher temperatures increase the danger of the occurrence of "bends," the eternal hazard of divers? If so, the diver would have to rise much more slowly to prevent the formation of tiny nitrogen bubbles in the body which cause intense pain and sometimes permanent brain damage. It appeared that if the body temperature did not rise, this would not be an added problem, but if it did, how much slower should the ascent be? No one knew, so it was fortunate that the swimmers' body temperatures actually rose little if at all.

Under what conditions could a safety line be used? Was the gear too bulky for proper maneuverability in tight holes? Although the diving team was undoubtedly as well prepared as any group could be under the circumstances, there was considerable tension evident as it pitched camp at the edge of the Hole on August 1, 1953.

Lines were rigged, and several hundred pounds of gear were laboriously carried down to the small rocky flat at the edge of the pool.

Aqualungs were filled with compressed air from the 285-cubic-foot high pressure tanks. Expedition leader Bill Brown put on his double-tank aqualung unit, face mask, foot fins, lamps, battery cases, depth gauge, underwater watch, and a few lead weights. Weighted down to this extent, he staggered clumsily in the shallows; as he slipped beneath the surface, he magically achieved the grace of a true aquatic dweller. He dropped down a few feet, testing the equipment. Satisfied, he gestured reassuringly toward the surface, turned, and leisurely fluttered downward out of sight into a world of distorted rock shapes looming grotesquely in the sapphire light.

Cautiously he made his way into a darker realm where his own lights would provide the only trace of illumination. His aqualung was functioning perfectly, and his depth gauge was showing ever-increasing figures as he passed below the range of visible light and into total blackness. Finally, without incident, he reached the 150-

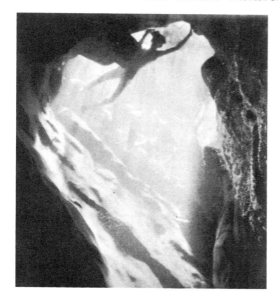

Swimmers descending into the depths of Devil's Hole Cave. *Robert Lorentz and Peter M. Neely*

foot level which was to mark the limit of his dive. Looking downward from the chockstone on which he sat, he saw no indication of the bottom, in spite of his powerful underwater light which had a range in clear water of nearly two hundred feet. Even today it is not known whether the depth of the Devil's Hole is measured in hundreds or in thousands of feet.

At this depth, the temperature of the water was little higher than near the surface. Dives to more than double Bill's depth are theoretically possible, but both the Southern California group and the National Park Service adhere to the philosophy that any record-breaking should be incidental to speleological investigation. So far, the divers have simply been too busy at lesser depths to continue on down.

If Bill had wished, he could have remained at the 150-foot level as long as his supply of compressed air lasted. But the longer he remained, the slower must be his ascent to permit the excess nitrogen to work out of his body without bubbling. Reluctantly pushing upward, he mentally catalogued the openings which appeared in the beam of his lamp. At this depth, the cave did indeed open up. At various lesser depths, several tunnels led in both directions along the axis of the pool, but their exploration would have to wait. Slowly, a tiny, deep-blue window appeared far above, then became aquamarine as he rose, ever-alert for the first sign of pain from the bends.

To the anxious crew above, the minutes dragged on, slower and slower. Other divers were prepared for instant submergence if assistance became necessary, and a fresh aqualung unit was ready if Bill showed signs of the bends and had to resubmerge to get rid of the deadly bubbles. Their watches showed that no unreasonable time had elapsed, however, and the continuous stream of bubbles from the depths was reassuring.

Suddenly, an excited chorus of shouts echoed in the narrow space. Far below, Bill's dim shape could be seen, ascending with slow, deliberate strokes. As he passed the more critical depths, his ascent quickened. Soon he broke the surface to report his discoveries to the excited crew. Whether this main interest was scientific or sporting, everyone there was exhilarated by the realization that this was the deepest American cave dive and the deepest dive into a hot spring ever accomplished. More important, this was just the beginning.

Late that evening, the exultant cavers descended to the small pool at the bottom of the nearby Devil's Hole Cave. A quick dive to a depth of forty feet was made to determine whether a passable opening existed which might prove to be an underwater connection between the two caves. No such opening was found.

Sunday morning, Brown slipped into the pool back at the Hole to begin the explorations. Almost immediately his bubbles vanished, for he found a passage extending back beneath the debris at the shallow end of the pool. It ended within thirty feet, so he turned his attention to the ever-tempting rear wall. Down and down it slanted into the enlarging fissure. Suddenly, when his gauge read eighty feet, the first important opening appeared, but it proved to be nothing but a narrow grotto—so narrow in fact that the swimmer had some trouble in turning around.

Undismayed, Bill followed the wall downward to the next opening. Here his gauge read 105 feet below the surface. Again, the opening was narrow, so that either his abdomen or the tanks on his back scraped the jagged wall as he eased himself upward through the narrow fissure. A few feet farther he discovered that just ahead there was a larger chamber, from which his light showed a wide tube leading far upward in the incredibly clear water.

He was not yet out of the fissure, however. As his aqualung again clanked against the wall, his air was suddenly cut off. Try as he might to breathe, he could not.

Immediately he realized what had happened. The control handle had struck a sharp rock which had turned it enough to shut off the flow of air. Quickly reaching backward, Bill turned the valve handle. Unfortunately, he failed to consider that he was indeed reaching backward, and turned it the wrong way. No air reached his lungs. A few more precious seconds were gone.

Slithering backward out of the fissure into the roomier main cavern, Bill twisted the aqualung unit around, inspected it, and turned on the other tank, all within an interminable thirty seconds. At last he began to breathe again, with hardly a mouthful of water to blow out of the exhaust valve.

Readjusting his gear, Bill reapproached the fifteen-foot fissure, this time with the tanks well clear of the wall. He passed through without incident, and the large passage began to slant upward. As he followed its course, the depth registered on his wrist became less and less. Yet no light appeared as he groped upward. Suddenly his head broke the surface of the waters and his lights illuminated the walls of a spacious chamber beyond the edge of the pool.

This was not the Devil's Hole Cave . . . The chamber was too large, and he had not traveled far enough before rising to the surface. This was a completely new discovery. Furthermore, at the end of the thirty-foot chamber, the dark opening of at least one passage led onward.

Was the air good? It seemed merely sultry, but the heat of the water and a contorted peak nearby indicated that there had once been subterranean volcanic activity here. Any toxic gases present in the water might very well have been trapped in the chamber. Bill therefore continued to use the dependable compressed air supply of his aqualung as he scrambled out for a hasty exploration. It was rough going; an aqualung diver's gear is not designed for climbing around on jagged limestone.

Bill calculated his reserve of air. It was probably fairly low, and the change of tanks had introduced a factor of uncertainty, so he slipped back into the water. Fortunately, there was little chance of losing the way, once the tight fissure was found. Even in an air-filled cave, it is difficult to maintain one's landmarks on an initial exploration, but here, the few side passages were no problem. As he passed through the fissure, the recollection of his earlier experience with it caused him some anxiety, but as he swam downward and beneath the ledge into the main passage, a reassuring faint blue glimmer reached him.

To the tired but exultant swimmer there was an urgency about getting up, but for proper decompression a slower ascent was necessary. Soon the reflected blueness brightened, and the tiny aquamarine window above came into view. Then he could accelerate his ascent until he was the target of a burst of excited questions from the supporting party.

A number of lessons were learned from this initial exploration and were applied in later dives in Devil's Hole and in underwater explorations in Lilburn and other California caves. The handle type of air control was abandoned. Use of a marker line, whereby a swimmer could find his way out in complete darkness, was adopted. It was decided that divers should never operate alone, that assistance should always be available. Just as in a "normal" cave, each person should be equipped with at least two dependable and independent sources of light. All divers should be fully trained in the technique whereby two swimmers may share one aqualung. All dives were to be timed and logged on the surface so that the divers might be notified if decompression was necessary, as it was in one case. Each diver was to wear a depth indicator and a special underwater watch was to be worn by one diver in each group. Finally, an extra aqualung set for emergency use must be ready at the surface during all diving operations.

Besides these general safety precautions, Simmons made special modifications of the aqualung equipment itself for maximum safety. The basic concept of the air supply is a 100 per cent reserve. In general use is a twin seventy-cubic-foot tank unit with individual valves. Each feeds into a manifold on which a two-stage Costeau regulator is mounted. When a dive is begun, the air from only one tank is used. When the swimmer feels the tell-tale pull of tank depletion, he turns on the other tank and begins his ascent, secure in the knowledge that he has as much air to ascend as he used to descend. If he wishes, he can shut off one tank when the two have reached equilibrium, and thus establish a known 25 per cent reserve.

Some of the divers have gone even further, carrying a second complete air supply: a small tank, regulator, and mouth hose. Plans for future, more extensive explorations call for the use of triple tank units, with reserve air supplies cached under water far inside the cave system. To date, as many as eleven dual tank units have been employed during operations at the Devil's Hole.

At these depths, lighting was at first a major problem. The early sets functioned fairly well, but collapsed under the excessive pressure. Now in use is an automobile six-volt sealed beam lamp powered by a nine-volt dry cell battery pack filled with paraffin. The bulb is mounted in a plastic handle with a grease-filled switch. A bright beam is thus obtained with minimal need for replacements. The battery pack lasts for five hours of continuous use, and no bulb has ever burned out under water. The second source of light is a rubber three-cell flashlight filled with mineral oil, with a metal clamp applied to minimize leakage. By the use of such precautions, the Southern California Chapter believes that it has reduced risk to an absolute minimum. Its members are determined that no serious accidents will mar their remarkable underwater record.

Calculations of the total horizontal distance Brown had traveled showed that although the completion of the passage to Devil's Hole Cave still appeared possible, it would be a much longer, more tortuous course than had been hoped. Even today, the passage has not been traversed, nor is it certain that it can be done without diving to unsafe depths.

The discovery of Brown's Room, however, was a remarkable feat from the standpoints of both science and exploration. Both phases of investigation were followed up. The air in the chamber was analyzed and found breathable, although the 91-degree temperature and 100 per cent humidity are excessive for any lengthy exertion. The problem of communication was greatly simplified by laying a telephone wire into the inner room. The job proved surprisingly easy; the free, graceful progress of the swimmers was scarcely hampered by the unrolling wire. With the advice and assistance of zoologists and students of the Scripps Institution of Oceanography, physical and biological studies are under way or planned at various depths and locations. Already, such finds as a new, apparently blind planarian, or flatworm, have resulted.

While the drama and excitement of the first few dives is largely past, further explorations are slowly advancing the known limits of the cave system. This, of course, does not mean that the safety precautions have made the dives a commonplace, hazardless affair. On one harrowing occasion, a diver felt the resistance in breathing that indicates an exhausted air supply, and had to come up without regard for decompression. Fortunately, his dive, like most in the Devil's Hole, was short, and he was none the worse for the experience.

Frank Leinhaupel and Roy Arnold carried caving gear into Brown's Room, and despite the uncomfortable heat and moisture, successfully explored the most inviting passage above water. They found it possible to climb, then descend gradually back down to a smaller pool about 150 feet away, now known as Leinhaupel's Pool. Frank and Roy returned via Brown's Room, but in 1954, Pete Neely emerged from the depths of this new pool by quite a different route. Investigating another opening just within the narrow fissure at the 105-foot level, with Richard Lawhorn and Robert Lorenz, Pete found a large, irregularly convoluted passage which not only led upward to the second pool, but also opened downward and toward the Devil's Hole Cave at the 120-foot level. Pete's passage thus appears to be the route which will be followed in future explorations. It promises extensive lengths of warm-water-filled cave to be explored, perhaps from both Devil's Hole and Devil's Hole Cave simultaneously.

Ultimately, sober scientific accounts of the remarkable contents of this cave system will be recorded, reducing adventure to matter-of-fact data and statistics. But the courage and skill of those explorers who first brought light into the depths of this spelean warm spring will not be forgotten.

Mammoth Cave's Underground Wilderness

One of the world's greatest cave systems, Mammoth Cave owes much to the insatiable curiosity of the men who daily lead parties over miles of tourist trails. At night and on their days off, the cave guides have climbed and ferried, squeezed and burrowed, their way into new chambers, to wonders never before seen by man.

Henry Lix belongs to the modern school of cave guides—university-trained and wearing the uniform and insignia of a Park Naturalist, United States National Park Service. Like many of the early guides at Mammoth, he began his cave exploring in early boyhood, but in his native Ozarks, rather than in Kentucky.

Lix has made an intensive study of the voluminous literature on Mammoth Cave. From his account of discoveries underground it is apparent that the cave has captured his imagination, as surely as it spurred on a succession of adventurous guides to penetrate farther and farther into its depths. His story is reprinted from The Living Wilderness, *where it appeared in celebration of the dedication of Mammoth Cave National Park, September 16, 1946. At present Lix is stationed at Great Smoky Mountains National Park.*

Henry W. Lix

The tens of thousands of hardy explorers who came over the Wilderness Road into "the dark and bloody ground" followed the trails of wild animals and of Indians through a forested wilderness. Every step forward created a new frontier. The first permanent settlement in what is now the State of Kentucky was established in 1774. By the turn of the century buffalo and elk had already disappeared from the state. With them went the true Kentucky wilderness.

Historic Mammoth Cave—another sort of wilderness—was discovered before the wilderness landscape of the Green River country had vanished entirely. In 1799 a black bear led a pursuing hunter into a large opening in a Green River bluff. This hunter, named Houchins, thus became the legendary discoverer of Mammoth Cave, a new underground frontier 360 feet under the surface, a new Kentucky wilderness which white men had never seen before.

The elements of uncertainty and mystery and danger of caves appealed to the pioneer's nature. Lives of many men became inseparably fixed to the great cave that came to be called "Mammoth" sometime before 1821. Traditions started. The cave "got in their blood." They were adventurous people, always going forward to see what was "just around the corner." This spirit of adventure in Mammoth Cave explorers has never waned. The exploration kept going on. It is going on today. But Mammoth Cave gives up its secrets slowly.

Nitrate deposits were discovered in Mammoth Cave before 1800. The dirt that had accumulated for ages on the dry floors of the cave contained large amounts of calcium nitrate, commonly known as saltpeter. Presumably this nitrate was derived from the droppings of bats that once lived in the cave in immense hordes. Many tons of saltpeter were produced in Mammoth Cave during the War of 1812, with the labor of slaves and oxen. Much of this nitrate was shipped to New Orleans to be used in the manufacture of gunpowder. Through a skillful combination of such things as Little Brown Bats, La Fitte the Pirate, Shreve's steamboat, "Old Hickory," and Mammoth Cave, America won the battle of New Orleans and the War of 1812.

As the men searched for new supplies of "peterdirt," they advanced farther and farther into the dark underworld. In this way they learned something of the apparently unending corridors of Mammoth Cave. The close of the War of 1812 brought an end to the highly profitable nitrate-leaching industrial period of the cave. In 1816 Mammoth Cave was opened to the public as a natural showplace and has been in continuous operation ever since.

Stephen Bishop, a slave, was the famous "first guide and explorer" of Mammoth Cave. He crossed the Bottomless Pit on a cedar sapling in 1837. The bridge opened up the extensive avenues beyond to exploration. It made possible the discovery of River Hall, Echo River, and Roaring River. Another obstacle had been surmounted by the explorer's impelling urge to go just a little farther.

For many years a big cave-in, which had probably occurred long before the cave's discovery, blocked all exploration in one of the five avenues converging from all directions at Grand Central Station. In 1923 an opening was finally made in this immense pile of fallen rocks. This opened up a major avenue and resulted in the discovery of one of the present outstanding exhibits of Mammoth Cave, the Frozen Niagara section. Here the explorers found "cave onyx" formations in greater abundance and beauty than any previously discovered in this cave. Immense Frozen Niagara itself, and the Drapery Room, and other nearby features, are spectacular. Until 1923 no one had ever seen Crystal Lake, as clear as only pure spring water can be. Its green is the green of clear deep waters, for Crystal Lake is thirty-eight feet deep. The water is caught and held in its downward underground course to Green River by a travertine dam 270 feet below the ground and 190 feet above Echo River and Green River.

And so the story of exploration goes on. The important discoveries in Mammoth Cave have been made by the cave guides. These men,

All parties of tourists visiting Mammoth Cave are conducted by well-informed official guides. Some of the men are third or fourth generation Mammoth Cave guides. *J. Wellington Young. National Park Concessions, Inc.*

who have assumed the responsibility for guiding the millions of visitors through the labyrinthine ways of the Mammoth Cave corridors, make up an unusual institution. They are proud of their profession and of their cave. It has become traditional for son to follow in the footsteps of father, even unto the fifth generation, and the entire guide group is the more closely knit because several of the families are related.

For a long time the cave guides speculated on where the enigmatic Roaring River might lead anyone daring enough to venture any distance beyond the Keyhole. Even before the Civil War, Stephen Bishop had gone far up the stream and reported that he had heard a "waterfall." In reality there is no waterfall on the river. The reverberations of the river waves lapping against the walls of the cave produce the roar of a waterfall, for the passageways are of the right size and contour to build up certain tones to an almost deafening crescendo. Perhaps the menace of Roaring River had a part in delaying the exploration of this subterranean stream. Yet, every time the guides went on a fishing expedition up the higher reaches of Roaring River they wished to venture further. The famous Mammoth Cave blindfish, *Typhlichthys subterraneus,* can be caught more easily in Roaring River than anywhere else in the cave; and so, once a year, and sometimes twice, the guides go up Roaring River to catch a supply of blindfish for display to visitors.

The roar of the river held a threat. There was a good reason why exploration had never been pressed on to the end, where possibly the waters might lap against stone roof and walls. Roaring River remained mysterious and forbidding. It is an undependable river, a treacherous river. Rains on the surface cause rapid rises. Sometimes the water rises as much as six feet an hour. At the Keyhole, Roaring River flows through a low-roofed channel, only about four feet from the ceiling. With heavy rains the river rises quickly at this point, and would thus trap anyone who had slipped through the Keyhole.

In 1938, more than a century after Stephen Bishop's crawl across the Bottomless Pit, two guides took up their gasoline lanterns and carried on the search. October had brought an end to the busy tourist season. Carl Hanson and Leo Hunt decided a bit of "caving" would be just the relief they needed from the monotony of routine guiding. The river was low that day, for October is a dry month, and the sky was clear. It would certainly not rain. Up the Roaring River they went. Up four stretches of water they paddled their flat-bottomed scow. Over three stretches of intervening mud and rock they drag-

Cave guides Leo Hunt and Pete Hanson take a "busman's holiday" by exploring new leads during periods of light tourist travel. *National Park Service*

ged the heavy scow, inch by inch. They came to a place where the river divided into two prongs. Here they stopped. Up above the muddy bank on the right they saw a hole in the wall. The opening was small and muddy, but it led gently upward and looked promising. Carl and Leo crawled into this hole. Slowly they crawled forward, pushing their lanterns ahead of them. To a cave explorer a new crawlway means a promising frontier. And so they squeezed and sweated and panted up the crawlway, over three hundred feet of fine wet sand. Then suddenly they were in an avenue with the ceiling ten feet high. This was encouraging—and a relief from cramped quarters. But their gasoline supply was running low, and they had to turn back. A cave is no place to be caught without a light.

At the guide house that evening the two explorers told M. L. Charlet, the cave manager, and their fellow guides about the new passageways. Everyone was excited. Would there be a magnificent cave at the end of the new crawlways? Or would they grow smaller and smaller till the men had to turn back? After a night's sleep the two men hurried off early to resume the exploration. This time they went on up the river beyond the muddy crawlway opening and found a better entrance to the passageway reached the previous day. This avenue led on for a mile, sometimes getting bigger, sometimes smaller. Then there was more crawling to do. The men crawled for a long way —to the end, where a large rock sealed the passageway. No, not the end, for a six-inch hole beneath the rock showed that the crawlway continued on the other side of the rock.

Back at headquarters again, Carl Hanson and Leo Hunt were very hopeful. The passageways which they had followed were leading towards Lee Ridge. And those who knew the Mammoth Cave region "knew" that someone would someday find a big cave under Lee Ridge. The area has good underground drainage. And the big caves are found under these ridges. Mr. Charlet agreed to let the men go out the following day with tools to make a way around the rock. But necessary precautions were outlined, for cave exploring is

Guides Carl Hanson and Leo Hunt had to drag their flat-bottomed scow inch by inch over stretches of mud and rock as they explored Roaring River. Water sometimes rises six feet an hour after a heavy rain, making exploration hazardous. *National Park Service*

dangerous business. Two other guides were sent along—Claude Hunt, a cousin of Lee's, and Carl Hanson's son Pete, who later gave his life for his country in the Aleutian campaigns of World War II. They were to help drag the heavy boats over the portages, help in working past the rock, and also act as liaison between the party and the outside world in case anything should go wrong.

The hole under the rock was quickly enlarged, and the men started forward once more. They had several hundred feet of crawling before the crawlway opened up into a passageway that averaged about ten feet high by fifteen feet wide. After working their way through an especially narrow part of this thousand-foot avenue, the four men were very tired. They dropped down on the floor of the cavern, which widened out at this point. Pete Hanson was a short distance ahead of the others, lying on his back, resting. He looked around him and then quietly stared almost straight above for a few moments. The ceiling glistened white and clean in the lantern light. This was something new.

Pete Hanson jumped up. "Come on, boys," he shouted, "we're in another world." Broadway in the old part of the cave was like this new avenue in size, but here was truly the "Great White Way." The clean, white, channeled walls and ceiling of the passageway sparkled as the light fell on their lining of gypsum crystals. The floor was blanketed with what looked like red pepper and salt, a mixture of red sand and gypsum crystals. It led on into the darkness like the smooth curves of a highway looming up before the headlights of a car at night. They hurried along.

This was beyond reality. This was the stuff of dreams. Snowy gypsum crystals encrusted the ceiling, the walls, and even the floor, for hundreds of feet. The party came to many crystal-covered grottoes in the walls and ceilings—some small, others large enough to walk into. Gypsum "flowers"—from tiny daisy-like blossoms to exotic "lilies" with curved petals up to fifteen inches long—grew from all sides in profusion.

And there were other wonders. Gypsum crystals occur in a wide variety of forms. The explorers found masses of gypsum that looked like balls of cotton or the cotton candy of carnivals. They found crystalline ribbons of gypsum spiraling from the walls like corkscrews, some of them eighteen inches long. In other places, round masses like snowballs studded the cavern walls. They came upon "pin-

Snowy cave flowers—crystals of gypsum—cover wall, floor, and ceiling for hundreds of feet in the New Discovery. *W. Ray Scott. National Park Concessions, Inc.*

cushions" bristling with gypsum needles. These transparent needles grow in clusters from floor and wall and ceiling; some of the needles are eighteen inches long and perhaps a thirty-second of an inch in diameter. As the men walked silently by, the crystal needles waved gently in the current of air. The men held their lanterns close, and the needles trembled in the convection currents produced by the heat.

They had been silent. But finally one of the men said: "Why, this is Paradise." And so a part of the main avenue where the gypsum formations are thickest has been named Paradise.

Travertine formations, the calcium carbonate dripstone and flowstone, are secondary in importance to gypsum in the New Discovery. But one of the most outstanding features of this section of the cave is a travertine dam forty-two feet long and over four feet high. The dam extends across the avenue from wall to wall. At one time it impounded spring waters to form a small subterranean lake, like Crystal Lake in the Frozen Niagara section of the cave. Here, by the dam, though, the lake dried up long ago when the springs that fed it failed, and fine mud cracks cover the floor that was once the bed of the lake. This travertine dam section, like several branches off the main avenue, contains beautiful small displays of stalactites, stalagmites, and cascades.

The four guides had much to see before the lanterns ran low. They almost ran over the smooth sandy floor of the cavern. Never before in the unknown ages that it took to dissolve and erode the caverns—in the ages during which mineral-laden moisture built up the layers of gypsum crystals and slowly laid down the travertine formations on walls and floor and ceiling—had the voice of man been lifted to echo and re-echo down these corridors. Carl and Pete Hanson and Claude and Leo Hunt made the first human footprints in the crunching sand, the first in all the cave's existence of possibly millions of years.

Only lower forms of animal life had passed this way before. The explorers found thousands of bat bones in dried-up little pools on the cave floor. Scattered bones discovered later show that the raccoon had been there, too. And perhaps a bear had wandered through those

Another of the endless varieties of cave flowers in Mammoth Cave. Some are as much as eighteen inches long. *Ross McDaniel. National Park Concessions, Inc.*

dark aisles. The men found several live bats hanging upside-down from the ceiling, asleep. In damp places cave crickets moved about in hordes, flickering their overgrown antennae. On the sandy floor in moist places were many beetle mounds. There was far too much to see.

Back down Roaring River and down Echo River the four exuberant men paddled their scow. They sang and shouted. No longer the quiet, reserved cave guides, they were returning conquerors. As they moved swiftly down the narrow winding channels of the treacherous stream, the air resounded with their songs—"Old Black Joe," "Swanee River," "One More River to Cross."

Mr. Charlet heard them coming. The men were safe and the news must be good. Mr. Charlet, then Cave Manager, was made Chief Guide when Mammoth Cave became a National Park. Deeply concerned with everything about Mammoth Cave—its guides, its welfare, its history—he listened with great satisfaction to the marvelous tale of the New Discovery. October 10, 1938. It was a memorable day for Mammoth Cave.

Superintendent R. Taylor Hoskins of Mammoth Cave National Park, Mr. Charlet, and all concerned with the management of Mammoth Cave immediately got busy making surveys to evaluate the discovery and to prepare the new part of the cave for exhibition to the public. With information from the surveys a point nearest the cave was located on the surface. At this place along the slope of a valley, engineers blasted a new entrance into the cave, to make it more easily accessible. By this time over four miles of passageways have been surveyed, and many additional miles of cavern and a new river are still being charted.

Kenneth Dearolf, biologist, made a study of the animal life in the New Discovery while it was still new and undisturbed. Specimens of bats and twelve species of invertebrate animals were found. Bones of bats, raccoons, martens, lynxes, and wood rats were also found. The bats and the remains of other vertebrate animals are an indication that the New Discovery had an outside opening. It is improbable that the animals entered by the devious way from Mammoth Cave's main entrance.

All those who have seen the New Discovery know how different it is from anything in the old part of Mammoth Cave. They agree that it has more unusual features, more abundant and spectacular cave formations. Impressive, too, is the whiteness, the cleanliness, of the passageways.

While the work goes on to prepare for the opening of the New Discovery to visitors, the cave guides will continue the exploration of this underground wilderness of Mammoth Cave. They will probe into the farthest corners of the New Discovery. They will paddle cautiously up and down its unexplored river. They will also look for a new crawlway, for one that might lead into an even bigger New Discovery. Under Joppa Ridge is a good place to search, perhaps, for that area too has good underground drainage to Green River. "Caving" is in the blood of the Mammoth Cave guides. They will continue to explore, to keep a sharp lookout for what's just around the corner in this subterranean frontier—mainly because they cannot help it.

Medicine, Miners and Mummies

There is no doubt that Mammoth Cave in Kentucky is the world's most famous cave. Dozens of books, here and abroad, have been written about it. Mammoth Cave was a sacred place to the original Indians. It has served its country in time of war and has been a mecca for visitors in time of peace. It has been used at various times as a chapel for weddings, as an auditorium for singers, as a tuberculosis sanitarium, as a station on the Underground Railway. It has been the scene of many thrilling stories of discovery. Mammoth has yielded valuable archaeological finds. It has contributed to the study of cave fauna and flora. One of this country's favorite national parks, it is known to every American.

HOWARD N. SLOANE

In 1954, scientists accompanied a National Speleological Society expedition to Mammoth Cave National Park to search for new antibiotic molds in the centuries-old dirt and sands in the caves of this area. Over a century before, in 1843, a colony of tubercular patients had lived in Mammoth Cave in an experimental attempt to cure pulmonary infection, literally living on the very substances which might have cured them one hundred years later. Though the 1954 scientists found no new molds, they did find that cave dirt contains the same molds from which present-day antibiotics have been developed.

The record of this remarkable early attempt to cure pulmonary infection begins in 1840 with a Dr. John Croghan of Louisville, Kentucky, a physician, author, and adventurer, and proprietor of the cave. At the time, the medical profession endorsed Dr. Croghan's attempt heartily. When the experiment proved a failure, his efforts were condemned for the very reasons for which they had originally been praised.

In the first volume of the *Western Journal of Medicine and Surgery*, published in Louisville in 1840, appears the following statement: "The enterprise is certainly novel and we see no reason why it should not, in many instances, prove more salutary than a visit either to north or south. Dr. Croghan, being a man of science, will understand in what manner to make arrangements for the accommodation of valetudinarians, while, with the taste of a gentleman, he will provide suitably for the comfort and amusement of those who in health, may visit this ancient resort of the extinct *people of Kentucky.*"

Actually as early as 1820, an unsuccessful attempt was made to induce the State of Kentucky to purchase Mammoth Cave and to

Remnant of stone cottage used to house tuberculous patients in Mammoth Cave in 1843. *J. Wellington Young. National Park Concessions, Inc.*

establish a hospital in one of its passages. It was not until 1843, however, that Dr. Croghan privately constructed twelve cottages for the housing of fifteen consumptive patients.

Ten of these huts were of wood frame construction and have since been removed, but two of them were built of stone and still stand. One of the stone houses was used as a dining room for the patients. Nine of the cottages were located in one area, thirty to one hundred feet apart. The tenth was isolated from the others. The two stone houses were built in the main avenue of Mammoth Cave, some distance away.

The cottages were apparently well furnished, with comfortable accommodations. Pitiful attempts were made by the occupants to grow plants and flowers around their cottage sites, but these attempts, like the hopes of the patients themselves, were doomed to failure.

One of the victims, a Mr. Mitchell, from South Carolina, lived in the cave for five months, after which he died and was buried in a cemetery near the cave. His body was later removed. Another patient subsequently died, and shortly thereafter all the remaining tenants left the cave, in much worse condition than when they had entered.

The medical records do not disclose the detailed case histories of these unfortunate people. There are, however, many interesting commentaries, which indicate the medical reasoning behind this novel experiment.

In 1850, Dan Drake, writing in *Principal Diseases of the Interior Valley of North America,* stated, in part: "The air of the cave has not been analyzed. Its sensible qualities are simply those of freshness. Dead animal matter does not become putrid but undergoes desiccation. There are no reptiles of any kind. Neither light nor sounds make their way into the deep recesses. They who have visited this great excavation, speak of wandering and clambering for a whole day without fatigue. They regard the atmosphere as invigorating. It may be that it holds saline substances in solution, which, entering the blood by lungs, favor its aeration, and thus ward off fatigue of exertion, or the mental excitement may support the strength of body.

"When saltpeter was manufactured there, it was observed that the health of the operatives was excellent, and that many 'ailing' or 'weakly' persons became sound in health and experienced increase of flesh. The oxen, also, that were employed, not only continued in good health, but became fat. With these facts before their eyes, the people near the cave have long believed that it might be made an advantageous abode for invalids, especially those affected with pulmonary diseases as they would escape all vicissitudes of temperature."

After commenting on Dr. Croghan's work, Drake continues: "But the results have not been encouraging. . . . The seclusion of a patient from the changes of the weather is not a positive influence, and by no means to be relied upon to arrest a malady which may occur independently of such vicissitudes. Then the solitude and silence, the darkness, the smoke, the atmospheric repose—for the wind is perceived only at the entrance—the want of exercise, the absence of many other excitors and sustainers of our mental and bodily activity, are counteracting agencies, not to be forgotten in a candid estimate. To render a sojourn in these subterranean cells effective in the removal of diseases, the patient should have occupations like those who once made saltpeter there; and to recommend such an abode to those who are too ill to labor and are in need of medication would seem injudicious, if not absurd. To what forms of chronic disease such a residence is in fact, best adapted, cannot, I think, be determined a priori. I would conjecture, however, that chronic bronchitis and functional disorders of the stomach, bowels, liver and spleen, would be more certainly relieved than any others. To these I would, conjecturally, add subacute opthalmia, obstinate ulcers, and other chronic affections of the skin. As to phthisis [tuberculosis] if the patient could engage in hard labor, and the tubercular transformation of the lungs had not advanced very far, it might, perhaps, be arrested but if he had reached the latter stages of the disease, he would do well to remain at home."

The medical history of Mammoth Cave did not end in the 1840's. On June 4, 1938, Dr. Nathaniel Kleitman of the Department of Physiology of the University of Chicago, assisted by Bruce Richardson, a postgraduate student, initiated an experiment in Mammoth Cave with interesting results. He set out to discover whether man could become accustomed to a cycle of life different from our standard twenty-four-hour day if the effects of sunrise and sunset were eliminated. The two subjects remained in the cave for five weeks, living on a twenty-eight-hour schedule instead of the normal twenty-four-hour one. Dr. Kleitman reported his experience in his book *Sleep and Wakefulness*.

The experimenters' reactions to the longer day were recorded principally by body-temperature measurements. It is a physiological fact that body activity throughout a twenty-four-hour day is reflected in temperature changes. Recorded at regular intervals, these fall into patterns, called diurnal curves. The average person, for example, sleeps when his temperature is lowest, usually 97 degrees or less, and is most active during the time when his temperature is highest, 98 to 99 degrees. This alternating rise and fall constitutes the diurnal rhythm. Collaborator Richardson was selected for the experiment because he had previously resisted a modification of his diurnal body-temperature rhythm.

The experiment had to be conducted away from interfering factors such as light, noise, and a higher environmental temperature in the daytime as compared to night. Mammoth Cave, which has a constant 54-degree temperature, met these conditions.

The two subjects moved into a chamber sixty feet wide and twenty-five feet high in a passage off the regular commercial route, about a quarter of a mile from the entrance. Beds, a table, chairs, a washstand, and platforms for recording instruments were provided by the management of the Mammoth Cave Hotel, which also furnished meals (once or twice daily) to be eaten at times determined by the experimenters' schedules, which varied each day.

The waking period was nineteen hours and the sleeping period nine hours. During their "day," the subjects were free to read, study, work on the records, or stroll through the cave. Their temperatures were taken every two hours during waking hours and every four hours during sleeping hours. Alarm clocks prevented oversleeping.

At the end of the first week of the five, Richardson became completely adapted to the odd routine, sleeping well and usually the full nine hours. For the last three weeks in the cave his temperature curves showed six waves per week, instead of the normal seven, indicating a complete adaptability to the twenty-eight-hour routine.

Dr. Kleitman, however, reacted differently. He slept well only when his sleeping hours coincided with his customary sleeping time and a low temperature level. At other times, he either had difficulty falling asleep or woke up too early, or both.

Furthermore, a comparison of the two subjects' charts showed that at certain times they were in opposite phase with respect to body-temperature. When this occurred, if Dr. Kleitman's temperature was at its maximum, and Mr. Richardson's was at its minimum, Richardson would have difficulty staying awake, whereas Dr. Kleitman was active, and vice versa. At some times, since one subject retained his normal seven-day rhythm, and the other adjusted to a six-day rhythm, the temperature curves were almost parallel, and the men were able to adjust more easily and work more compatibly.

After he left the cave, Richardson's temperature quickly became readjusted, and in three weeks he was back to normal. Dr. Kleitman's, however, took longer to return to normal. The experiment explains why some people are able to adjust easily to a routine of working at night and sleeping in the daytime and others are not. Apparently, people who are required to work odd shifts, or at night, might well consider taking tests to determine if they have the adaptability to make these changes.

Dr. Kleitman concluded from the results of the experiment that "there is no foundation for assuming that some cosmic forces determined the diurnal cycle of function aside from rest, movement, food intake, and sleep."

Just as Mammoth Cave has been the "testing laboratory" for the physiologist, and the "clinic" for the physician, it has served as "laboratory-factory" for the chemist.

Nitrate, used in explosives and gunpowder, had not been produced on any large scale prior to 1812. With the coming of war, it became as essential in those days as uranium ore would be today; in fact, the situation was somewhat comparable to the present-day development of uranium ore deposits.

Searches for sources of supply for nitrate were extensive, and the most accessible supply was found to be in cave deposits of saltpeter. During the War of 1812, many caves were mined for saltpeter, although few of them had the large working area of Mammoth Cave. Much saltpeter was taken from the Shenandoah Valley caves, but probably the largest quantities came from the cave earth in Kentucky, it being said that some 400,000 pounds were produced in Mammoth Cave alone. The cost of this production was exceedingly high, and after the war, when imported nitrates were again cheaper than cave saltpeter, most of the operations were discontinued.

That Mammoth Cave had long been a favored abode of bats

may account for the abundance of the nitrate deposits. Today there are relatively few bats in Mammoth Cave, but in the early 1800's, at least, they were present in great numbers.

Two large groups of leaching vats to refine the saltpeter were built in the cave during the War of 1812, each holding approximately one hundred bushels of niter-containing dirt. Extensive pipe-

Wooden pipes in Mammoth Cave which carried water to saltpeter leaching vats during War of 1812. *W. Ray Scott. National Park Concessions, Inc.*

lines carried water from outside the cave to the leaching vats within. The pipes were made by boring two-inch holes through logs chosen from the straightest tulip or oak trees. One end of each log was sharpened, very much as a pencil is sharpened, and the hollow point was inserted in the two-inch hole of the succeeding log, thus forming a continuous pipeline. These pipes are so well preserved that they can be seen in the cave today.

Most of the work was done by slaves, who carried the earth in bags to oxcarts which were then hauled to the leaching vats. Water from the pipes was poured over the earth in the vats, dissolving out most of the saltpeter. After the solution flowed down into a collecting tank, it was pumped to a higher tank from which it was carted out or from which gravity caused it to flow to the entrance of the cave.

The "false saltpeter," as it was called, had to be converted into true saltpeter, or potassium nitrate, before it could be used to manufacture gunpowder. To sum up a complicated operation, this was done by heating the solution to the boiling point, then allowing it to cool, at which time the saltpeter crystals would settle in troughs. The crystals were carried out in buckets and shipped to gunpowder plants.

Early visitors to Mammoth Cave did not, of course, have the benefit of electricity. The inadequacy of kerosene lamps for lighting the tremendous passages led to the development of an art of flare-throwing by the guides which was astounding in its accuracy. Each

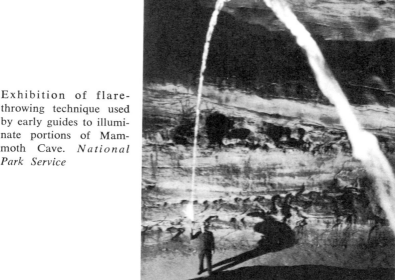

Exhibition of flare-throwing technique used by early guides to illuminate portions of Mammoth Cave. *National Park Service*

guide made his own flares of kerosene-soaked rags. With the aid of a flare-throwing stick, burning rags could be hurled for distances of over one hundred feet with such skill that they would land in small nooks and crannies, where they created unique lighting effects. Many inaccessible spots in the cave are blackened by the carbon from the burning rags. A few of the present-day cave guides still demonstrate this flare-throwing technique.

The first guide and explorer of Mammoth Cave was Stephen Bishop, a slave, believed to be half Negro and half Indian. He somehow acquired a knowledge of geology and other sciences as they related to the cave and even had a smattering of Latin and Greek. It was through Stephen Bishop's efforts that the famous Bottomless Pit was first crossed in 1837, making possible the discovery of Echo River and other sections.

Today many of the guides in Mammoth Cave are descendants of the pioneers who helped in its development. Indeed, most of the exploration of the cave has been done by guides, including the New Discovery made in 1938 by Carl Hanson and Leo Hunt, and described in the preceding chapter by Henry W. Lix.

The history of Mammoth Cave can be traced back five or six hundred years to a Pre-Columbian Indian tribe which used the cave extensively. Some Indians apparently remained near the cave entrance, but relics or artifacts were also found deep in the cave.

The constant temperature and humidity, and possibly the presence of saltpeter and other salts, have helped preserve not only the artifacts but several Indian mummies, the first of which was discovered in 1813. All the mummies were well preserved through desiccation.

In 1935 Grover Campbell and Lyman Cutliff, two of the Mammoth Cave guides, were crawling on a high ledge, two miles from the entrance. Their curiosity had been aroused by the odd scratch marks found on the gypsum-covered walls in many parts of the cave, and they were hopeful of making a major discovery. Squeezing through a narrow passageway between large rocks, the two men were astonished to find that Campbell's hand had come to rest on the head of a mummy. The body was partly hidden beneath a huge slab of rock.

Recognizing the importance of their discovery, the explorers excitedly retraced their steps and reported their findings. The director of the National Park Service in Washington, D.C., was advised promptly, and Dr. Alonzo W. Pond, who was then engaged in archaeological research at Jamestown Island, Virginia, was rushed to the scene.

It is a tribute to the intelligence of Campbell and Cutliff that when Dr. Pond arrived, nothing had been disturbed. The thrill of the discovery is best told in Dr. Pond's own words, which appeared in *Natural History* in 1937:

"Many years of archaeological exploration on four continents had given me more than my share of 'firsts,' but nothing gave me the thrill I experienced as I sat on that narrow ledge in Mammoth Cave with the discoverers and saw with my own eyes the perfectly preserved body of that prehistoric miner trapped at his work centuries ago.

"Nothing had been disturbed. The ledge was covered with loose, dry sand over which had settled fine, black soot from the torches of ancient and modern 'cavers.' In the tragic tableau before us, time had stopped centuries ago. With the event of death and the subsequent drying of the man's body the scene had remained unchanged. Here was preserved one of the most complete chapters in the life of prehistoric peoples. . . .

"As I sat there gazing at the dried body of the prehistoric man a host of questions clamored to be answered. For what was he looking? What was so precious that he would dare penetrate the darkness and brave the mystery of the cavern to secure it? Was anyone with him at the time of the accident? What tools did he use? What food did he eat? At what season did he come to the cave? How many centuries had passed since the flickering light of his torch ceased to throw weird shadows on the cave walls? Long I pondered those and many other questions. Then the archaeologist in me stirred and I set to work methodically to study the clues for answers. It is usually difficult to solve a detective case like this where the trail has been cold for centuries; but in the end we had the answer to nearly every question."

As the story unfolded, it seemed that the ancient explorer had been trapped, face down, under a block of limestone weighing between six and seven tons, but that part of his body was exposed, indicating that perhaps the victim had not been instantly killed by the rock fall but had died a lingering death. Careful examination of the surrounding sand disclosed two bundles of oak sticks tied with grass, a small piece of a gourd, a hickory nut, a bundle of reed, and some pieces of excreta.

For several weeks and with great success, Dr. Pond searched the accident site and many side passages for further evidence of long-ago visitors. Dr. Pond and his assistants discovered paths marked by pieces of burned reed torches; they found trails worn smooth from use; they located sticks worn down as if used for digging; and they found mussel shells with worn edges. Yet none of these told why

Tracks left by prehistoric Indian gypsum miners in the dust of Barefoot Ledge, two miles within Mammoth Cave, Kentucky. *National Park Service*

their users had roamed through the cave. One question, however, constantly drummed on Dr. Pond's mind. Why, wherever any evidence of human occupation was found, were the gypsum-covered walls scarred and scratched? Why had all reachable crevices, ledges, and crawlways been stripped of their gypsum? Why were there so many blunt-edged sticks, worn-down pieces of stone, and mussel shells? It was evident that it was the gypsum that had drawn the visitors deep into the cave. To what use had the early miners put this material? Had it medicinal value? Was it used as a love potion? Or was it used to make white paint?

Many other artifacts were found in the search for a solution to the riddle: many types of sandals, usually woven from the bark of elm trees, several ladders, some pieces of tree trunks, which had been used as steps, stems of wild grapes, gourds, and sunflower

The six-ton tomb rock was bound in a metal cradle and lifted off the mummy pinned beneath it. *National Park Service*

seeds. The plant material indicated that the visitors worked in the cave during the autumn, when these plants were ripe. At last, as if to prove the entire theory, a store of gypsum crystals was discovered carefully hidden under the rocks.

There was still much to be learned about "Lost John," as the mummy was fancifully named by the researchers. Was he Indian, and if so, why were there no bones of deer, bear, or bison, no arrow points or axes, no evidence of pottery, no beads, corn, or tobacco, usually associated with Indians?

Finally, Park Service officials authorized the release of the mummy from under the boulder for more detailed scientific studies. Getting "Lost John" out was a major task, necessitating the erection of a thirty-foot tower, from which were suspended chain hoists to lift the more than six-ton tomb rock.

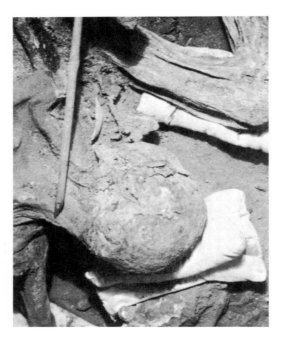

Close-up of "Lost John" following the removal of the tomb rock and the excavation of the accumulated sand around the body. *National Park Service*

At last, "Lost John" was freed, and the complete story was interpreted as follows by Dr. Pond, in his 1937 *Natural History* article:

"We saw what lay beneath the rock and knew the story of what had happened there so many centuries ago. . . .

"Through the flickering lights and shadows of a reed torch, an old Indian gypsum miner, barefooted, moved along a narrow ledge high above the floor of the cave. Carefully he laid two bundles of oak sticks in a niche in the rocks, which he would ignite later to help dispel the darkness and drive away the spirits. With similar purpose this man of destiny placed a long bundle of dry reeds close by on the steep slope of that treacherous ledge. He sat down to munch his meager meal of hickory nuts. Curious shadows leaped about the walls of the cave as the torch light wavered. A stone rattled down the ledge momentarily shattering the awful silence as it crashed on the floor below.

"His lunch finished, the old man squatted silently a moment. The darkness and the silence of the cavern brought him close to a sense of the eternal mystery. Fear of the darkness, awe in the presence of the mysterious unknown, were overmastered by his intense religious fervor and the desire to secure the sacred gypsum. He touched his polished clam shell amulet.

"For two miles he had wandered through the majestic passages of the great cavern, a puny soul wandering in eternity, his footsteps guided by the fitful, yellow glare of a reed torch. Through tortuous rock falls he had clambered where even his precious bundle of torches was a burden. His faith in the sacredness of the tribal need for gypsum, precious ceremonial paint of the ancestors, had driven him on through the silence.

"At last he was ready. He adjusted the folds of his fiber blanket across his hips, knotted it in front and drew a large part of it over his chest, bib-like. Cautiously he crawled under a great block of limestone that had lain for centuries on the steep ledge. He knelt on the loose sand of the ledge. With a large chunk of limestone he started to chip away the gypsum. His position was cramped, awkward. He moved his left foot for greater comfort; it dislodged a small key stone beneath that huge block of limestone!

"An agonized scream of terror shattered the cavern's stillness! A few pebbles rattled down over the ledge to the cave floor. The reed fire flared up and flickered out. Silence and the blackness of eternity descended again on the great cave. Death had posed a tableau of prehistoric man's intimate daily life. The strange chemistry of the cave began the process of preservation. A rat gnawed a little of the body, then left its job unfinished. Other gypsum miners worked on a higher ledge. Sand from their diggings trickled down with hour glass slowness to bury the ledge of tragedy. Eternal minutes in the cavern grew to years. Decades lengthened into centuries."

The Valley of Virginia

Being so close to the metropolitan centers of the East, the caves of the Valley of Virginia have doubtless introduced more people to the marvels of the underworld than any other cave region. Here are caverns with superlative scenic attractions and rich historical associations.

Through the years hundreds of excursion trains, thousands of sight-seeing buses, and hundreds of thousands of automobiles have poured visitors into the valley for cave tours—into a valley whose caves knew the step of George Washington and Thomas Jefferson. These are the caves which, decades later, served temporarily as barracks, sometimes for troops of Stonewall Jackson, sometimes, as the fortunes of war varied, for Union forces from Ohio.

William E. Davies knows these Appalachian valleys well. In the course of preparing his Caverns of West Virginia, *he explored more than four hundred caves in that state. He visited hundreds more in Virginia and Tennessee, and every known cave in Maryland for the official cave publication on that state. In Pennsylvania he explored caves for the State Geological Survey. His numerous published studies on the Appalachian caves have helped clarify scientific thinking on the origin and development of caves.*

WILLIAM E. DAVIES

To many people, the Valley of Virginia is synonymous with "caves." To the tourist, it brings to mind beautiful commercial caverns that have long topped the sight-seeing lists; to the explorer or spelunker, it conjures up visions of thousands of square miles of limestone country, honeycombed with intriguing caves and pits. This valley is a vast lowland along the eastern part of the Appalachians from Pennsylvania to Tennessee. In Pennsylvania, it is the Chambersburg Valley; in Maryland, the Hagerstown Valley; in Virginia, the Shenandoah Valley; and in Tennessee, the Holston Valley.

Regardless of differences in names, the features of the Valley of Virginia are the same. It is a country of rolling hills and numerous shallow stream valleys. On either side, the big valley is bounded by the long, high ridges of the Appalachians, forest-covered mountains of hazy blue, contrasting sharply with the rich green and brown tones of valley farmlands. Beneath the valley is a series of crumpled and tightly folded limestones, several thousand feet thick. And here have been carved out the many caves for which the region is famous.

Many caves within the Valley of Virginia are open to the public. Every one of them is outstanding and well worth a tour of inspection. Their names are household words: Battlefield Crystal, Dixie, Endless, Grand, Luray, Massanutten, Melrose, Natural Tunnel and Chasm, Shenandoah, and Skyline. But let us take a look at a few of these. Grand Caverns is located at a town appropriately named Grottoes, about fifteen miles northeast of Staunton. The cave is an attraction both to the tourist and to the geologist, for along its passageways are formations of exquisite beauty, and its geologic features are of interest to all who are curious about the origin of caves.

Grand Caverns was one of the first caves found in the valley, having been discovered in February, 1804, when Bernard Weyer dug into the entrance while trying to recover a trap. Known for over a century as Weyer's Cave, it was opened to the public not long after its discovery, and Thomas Jefferson was a frequent visitor. During the nineteenth century, both the Baltimore and Ohio and the Norfolk and Western railroads brought hundreds of tourists there annually, and on the former, a station and town bore the name Weyer's Cave.

Like all the commercial caves of the valley, Grand Caverns is rich in beautiful formations. The most unusual are the "shields," flat slabs of calcite crystals that grow at angles of 45 degrees or more from the wall with no apparent support. Their origin has long perplexed geologists and mineralogists, and as yet no satisfactory explanation has been advanced. Unique in size and perfection is the Bridal Veil shield, eight feet in diameter, standing at a steep angle from which long curtains of calcite fall like folds of a Greek robe. The total height of the formation is fifteen feet.

The high, broad passages in the cave run parallel to the beds of

The Bridal Veil shield formation in Grand Caverns, Virginia. The circular portion is eight feet in diameter, and the entire formation is fifteen feet high. *Virginia Chamber of Commerce*

rock for short distances and then change direction to cut directly across the tilted and arched rocks, apparently without relation to any specific bed of limestone. Instead, the cave follows two sets of prominently displayed joints in the rocks.

Like many of the Valley of Virginia caves, Grand Caverns was occupied by troops at various times during the Civil War. Some served under Stonewall Jackson, whose name is carved on the cavern walls.

Another cave, near Grand Caverns, is Fountain Cave. A showplace during the last century, but now closed, it is a veritable museum of nineteenth-century commercial cave operations, for in it are many of the old fixtures for lighting, as well as old pathways and signs. Madison Cave, a tourist attraction over a century ago, but since closed, is reputed to have been visited by George Washington, who inscribed his name on the walls. (Washington's signature, with the date 1748, is also carved on the walls of a cave near Charlestown, West Virginia.)

Skyline Caverns, just south of Front Royal, Virginia, has within its deep recesses some of the most delicate cave formations known. Well back in the cave, near the end of the public tour, is a circular passage that is part natural and part man-made. This passage connects several small rooms that contain clusters of anthodites, or cave "flowers," on their ceilings, the prize feature of the cave.

These small rooms had been sealed off from the rest of the cave by clay fills. Now the visitor passes through a double set of doors on entering and leaving the passage. The doors help to maintain the uniform conditions of temperature and humidity necessary for the preservation of these delicate "flowers." The anthodites are radiating clusters of thin needles of calcite, the needles pure white in color and up to four inches long. Their origin has not been satisfactorily explained, although it is apparent that the high humidity of the sealed-off chambers was an important factor to their development.

Tucked away at the head of a deep, narrow valley, some four miles south of Ronceverte, West Virginia, is Organ Cave. It, too, is open to the public. Organ Cave was one of the first caves in America to attract world-wide attention. Near the end of the eighteenth century, saltpeter miners, while removing the cave earth, came across the fossil bones of a prehistoric sloth. Eventually news of this find reached Thomas Jefferson, who procured the bones and deposited them with the American Philosophical Society. They were studied and described by Jefferson and other scientists, both in American and foreign journals. Unfortunately, Jefferson did not specifically locate the cave in which the bones were found, and for over a hundred years, scientists and cave explorers played a game of detective, evaluating each scanty clue to its identity. In 1949, more fossil bones, those of a peccary, were discovered in a remote and forgotten part of the cave.

It was in 1949, also, that Organ Cave was discovered to be one of the largest caverns in the eastern United States. In that year, a party of explorers discovered several miles of hitherto unknown passages. Actually, the party had not intended to visit Organ Cave at all but were investigating another cave, Hedricks, whose entrance is five miles to the north. Taking a peek into a small side passage of Hedricks the explorers soon found that it led to a vast system of rooms and broad passages, which they followed well over six miles. At the end of the trek, they found themselves in Organ Cave, in a part which bore several signatures dated over a century ago.

Perhaps the major attraction of the cave is in a low, long room near the end of the usual tour. Here time has stood still. Surrounding the visitors are reminders of the saltpeter mining carried

Peggy Mueller of the National Speleological Society examining saltpeter hoppers in Organ Cave, West Virginia. *Charles E. Mohr*

on during the Civil War. Hoppers and troughs occupy most of the room. The hoppers are filled with earth and are in such an excellent state of preservation that the visitor feels the miners have just stepped out for lunch and will soon return.

Down through much of American history—the Revolutionary War, the War of 1812, the Mexican War, and the Civil War—the "saltpeter monkey," as a miner was called, was an important part of the wartime economy. He worked long, hard days in the dim recesses of caverns lit by the smoky, orange light from many pine fagots stuck in the walls or floor of the cave. His tools during wartime, especially in the Confederacy, were simple and were made of wood—metal being too precious for use except where there could be no substitute. Many of the relics from saltpeter mining are in caves used from 1861 to the end of the war. The Confederacy was cut off early in the war from the usual sources of nitrate and

Winch used by saltpeter miners during Civil War, found in Haynes Cave, West Virginia. *William E. Davies*

had to seek it in caves, barnyards, and special nitrate-developing plants run in conjunction with powder mills. Caves supplying the vital nitrates are found throughout the southern states, but the most important were in Virginia, West Virginia, Tennessee, Alabama, Georgia, and Texas.

Most saltpeter caves are relatively dry, and conditions in general favor the preservation of relics. Hoppers and troughs, however, do not always fare well, as the earth left in them hastens their decay. Puddling boards or paddles, ladders, buckets or vats, and burned fagots have been recovered in quantity. Mattocks, shovels, knives, and personal equipment have also been found.

More interesting than the equipment are the graphic records left in some of the caves by the miners. In many caves the pick marks in banks of saltpeter earth remain as fresh as the day they were made. In Greenville Saltpeter Cave, in southeastern West Virginia, there are cart tracks, imprints of mule shoes, and human footprints in the clay floor, looking as though they had been made but a day or two before.

Less common are the names smoked or scratched on the walls by the miners. Most of them are the names of local people, and their descendants are to be found in those localities today. In addition to the marks of the saltpeter miners, there are the signatures and messages of thousands of visitors who have explored underground.

Indians commonly left records of their visits to caves in the form of pictographs or petroglyphs. In a small shelter cave in sandstone, several miles above Clarksburg, West Virginia, are several well-executed pictographs. Figures of dogs, foxes, sunfish, rattlesnakes, and human hands are etched life-size on the walls. Faint traces of pigments added to the pictures still remain. Unfortunately, some of the drawings have been defaced by later artists—surveyors who left bench marks painted over the original pictographs.

Melrose Caverns, often known as the Blue Grottoes, contributes a unique bit of written history to the story of the Civil War. The caverns are located a few miles north of Harrisonburg in the heart of the Valley of Virginia, a battleground during four years of the war. They were used as shelter by both Confederate and Union troops, as the valley changed hands in succeeding campaigns. Each occupation of the cave can be traced by the names and dates carved or smoked on the walls and formations.

The cave was known long before the war, having been opened for inspection in 1824 (dates inscribed on the walls go back to 1811), but it received little attention until 1862, when Frémont camped in the pear orchard near its entrance before the Battle of Cross Keys. In the main rooms of the cave, the Record Room, Frémont Room, Century Hall, and Black Mountain, are hundreds of inscriptions of the names of Union soldiers. Many of them are from Ohio—C. Seewald, Tiffin, Ohio, C. D. Bergin, Co. B, 4th Ohio, and W. B. Mc-

Guire, Co. B, are but a few of those carved boldly on the Registry Column in the Frémont Room. In one room, a replica of the United States shield has been carved on a stalactite.

The Registry Column in the Frémont Room of Melrose Caverns, Virginia, showing the carved names of Civil War Union soldiers. *Burton S. Faust*

Shortly after the Federal troops withdrew from the area, Stonewall Jackson's men, who had been camped near Weyer's Cave to the south, occupied Melrose Caverns, scratched out many of the Federal names, and superscribed their own. When Federal troops under Sheridan carried out their raid on the Valley of Virginia, they again occupied the cave and many names bear dates of 1864.

Other evidences of the occupation by troops are a huge bullet-scarred column that served as a target and a defaced stalagmite, the Scarred Veteran, that served as a source for souvenirs.

In limestone areas the visitor can sometimes go in one side of a mountain and come out the opposite side. One of the most celebrated of these tunnel caves is the Sinks of Gandy Creek in central

West Virginia. Gandy Creek rises in Randolph County and flows northward through the vast Monongahela National Forest. Once cut for lumber, the forest is rapidly reverting to its original wild state, and the few people who live in the area are almost as isolated as in pioneer days. There are no paved roads in the forest, and what dirt roads there are generally follow the meanderings of old logging railroads.

Four miles from its head, near the settlement of Osceola, Gandy Creek flows across a broad, swampy meadow, and to the casual observer it appears that this is the end of the line. A broad spur of Yokum Knob stretches across the valley, joining the lower flanks of Allegheny Mountain. But Gandy Creek, being a real Mountaineer State stream, is not to be stopped by a mere ridge, so not being able to flow over, it flows under it.

Porte Crayon's classic words describe the entrance to the sinks in terms that are hard to improve upon: "Looking up the glen, the

Looking out of the entrance of Sinks of Gandy Creek, West Virginia. *William E. Davies*

vista is bright as fairy-land, ending with a distant glimpse of blue hills. Turning down stream, a grim, menacing cliff rises square athwart the glen, closing it suddenly and shocking you with its unexpected propinquity. At its base is an arched opening fifty feet wide by about twenty in height—a gaping mouth which swallows the little river at a gulp. There is no gurgling nor choking, but the stream glides in gently and lovingly. . . .

"Wading in some forty or fifty yards, we find the subterranean stream still smooth and practicable, without any roaring or other indication of an interruption in its current. But its winding course soon shuts out the daylight, and as we had no torches no attempts were made to push our explorations further."

In that darkness into which Porte Crayon gazed at the end of his explorations lie 3,000 feet of rugged cave. For those who have penetrated to the lower portal, the traverse is mostly in the stream. Here and there, deep pools or narrow constricted channels force the explorer to leave the stream and clamber along ledges or gravel banks. Fortunately, the cave is large throughout with the main passage at least six feet high and thirty to a hundred feet wide. The exit at the north end of the sink is a little tricky, for, though daylight can be seen through low, water-filled openings, the real exit is concealed by an offset in the passage. This exit opens into a densely forested cliff, and Gandy Creek, on reaching the opening, resumes its normal surface course, flowing north to join the Cheat River, the waters of which ultimately reach the Ohio.

There are many other caves similar to the Sinks of Gandy Creek. Most of them are routes for subterranean streams, which at times bring in debris that blocks the caves. In the Sinks of Sinking Creek in southeastern West Virginia, whole trees, huge logs, and other driftwood have piled up in the cave until they form a timber dam forty feet high. In the Great Saltpeter Cave in east-central Kentucky, one can walk dry-shod through a hill, for unlike most tunnel caves, this one has no stream.

One cave in southwestern Virginia is not only a tunnel in form but actually serves as a tunnel for the passage of the Appalachian Division of the Southern Railway. The tunnel is 1,557 feet long and cuts through a ridge that stands 400 to 750 feet above the tunnel floor. The tunnel averages 130 feet wide and about 100 feet high. Mid-

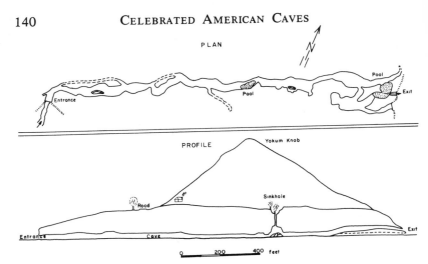

Map and cross-section of Sinks of Gandy Creek, West Virginia. The cave runs completely under the mountain and comes out on the opposite side.

way, it becomes a vast opening in the form of an amphitheater that is 600 feet in diameter, with vertical sides 400 feet high.

The tunnel was well known in pioneer days and has been a mecca for tourists for almost a century. The railway was constructed in 1882 to tap the coal fields in southwestern Virginia. The original "contractor" for the tunnel was Stony Creek, which for untold centuries has been busy enlarging and altering an ancient cave that it usurped for its course. Today, Stony Creek is a small stream that shares the giant passage with the railroad, telegraph lines, and tourists.

A cave that almost became a railroad tunnel is Jewell Cave, West Virginia. The entrance is along the Chesapeake and Ohio Railroad, on the Greenbrier River near Ronceverte. In this region the Greenbrier flows in a meandering, sinuous course. The railroad originally followed the south bank of the stream, which added considerably to its mileage, but the longer loops have gradually been cut across by tunneling, until today the railroad runs in what is almost a straight line.

About a generation ago railroad engineers considered cutting through the loop at Jewell Cave. It appeared that the cave offered great possibilities as a tunnel. Its main passage is relatively straight, with a width of fifteen feet or more, and is about twenty-five feet high. It extends about a thousand feet in a southeasterly direction which is approximately a third of the distance across the loop. Surveys were run through the cave by the railroad, and the survey numbers, embossed in red paint, are still evident on the walls and ceiling. Even the officers and directors of the railroad visited Jewell Cave. However, when time for construction came, the idea was abandoned.

Less well known but equally interesting are the pit caves in the Germany Valley in central West Virginia. Here a broad arch of limestone floors a high, rolling valley enclosed by the lofty ranges of the Alleghenies. Within the limestone are several caves noted for the depth of pits that intersect their otherwise horizontal passages.

Most renowned is Schoolhouse Cave, which has been the site of numerous expeditions—"mountain climbing underground" such as described by Ida V. Sawtelle elsewhere in this book. The cave is most noted for a series of pits opening in the floor and extending downwards as many as three hundred feet. The pits, sometimes singly, sometimes in pairs, cut the passage at several points, and traversing the cave is a succession of mountain-climbing events, first descending one side of a pit and then climbing out the other, over and over again.

Close by Schoolhouse is Hellhole. This cave opens at the base of two small, steep-sided, and very dangerous sinks, which funnel down into the ceiling of a tremendous room. Between the bottom of one of the sinks and the floor of the room below is open space equal in height to an eighteen-story building, 180 feet high and 150 feet in diameter. From the room several passages lead off in different directions, and along these there are several deep pits connecting with lower levels, the ultimate depth of which is 425 feet below the surface.

Caves of the type of Schoolhouse and Hellhole are for the rock climber and other spelunkers well versed in the skills of mountaineering in the dark. But the Appalachians have so much to offer the casual tourist and the history-conscious explorer that no one need fret at passing up the perilous pits of Germany Valley.

Carlsbad Caverns

Although it has been open to the public for less than thirty years, Carlsbad has won a world-wide reputation as one of the scenic and geological wonders of America. It is one of the gems of our National Park System. Within the far-flung boundaries of Carlsbad Caverns National Park are many scores of caves, among which New Cave and others in Slaughter Canyon are especially notable. They doubtless will provide scientists with rich hunting-grounds, once the Park Service has adequate personnel to supervise or carry out the necessary exploration and study.

T. Homer Black was born in Missouri, but never gave caves much thought, he says, until he was stationed at Carlsbad as Park Naturalist. In telling the story of this underground fairyland, he writes from twelve years of experience.

T. Homer Black

Carlsbad Caverns, in the dry foothills of the rugged Guadalupe Mountains of southeastern New Mexico, has been noted since the early twenties for its incredible extent and for the magnificence of its decorations. The caverns consist of a twenty-three-mile labyrinth of passageways, huge corridors, and great chambers, with a record vertical range of 1,076 feet. Dr. Willis T. Lee, the first scientific explorer of Carlsbad, described it as "the most spectacular of underground wonders in America. For spacious chambers, for variety and beauty of multitudinous natural decorations and for general scenic quality, it is king of its kind."

Situated, like all major caverns, in a limestone terrain, Carlsbad owes its existence to the corrosive action of slowly moving ground water on the limestone. Since iron oxide is the primary impurity, the formations are colored in exquisite pastel tones, ranging from faint, creamy tans through warm peach to rusty reds. Many of the formations are of majestic size, and although the color is not as vivid or as varied as that in some other famous caves, the impression is one of quiet, serene grandeur. In reality, considerable color is present, but it has been dimmed by the dry, opaque scale which covers most of the formations. Carlsbad is now inactive, or dry, and has been so for thousands of years.

The enormous size of the caverns' rooms and galleries has been made possible by the peculiarities of the massive structure in which it formed. Carlsbad Caverns developed within a gigantic reef laid down in a Permian sea which covered a large portion of western America a little more than 200,000,000 years ago. Ancient land forms, inherited from an older geologic period, dictated the areas in which this reef could grow. Since it developed, in the main, through the carbonate-precipitating activities of algae, the primary limiting factor in its growth was the depth to which sunlight could penetrate in sufficient amounts to allow photosynthesis to be carried on. This was, of course, determined by the conformation of the sea bed, and hence it was on the banks created by the drowned highlands of the older land surface that the reef began to form.

Here, on the edge of the Delaware Basin, where nutrient-rich waters rose from the deeps, life flourished along the submarine slope far enough from the shore to escape being buried and killed by sediments brought in from the surrounding continental area. Calcareous sponges seem to have served as a reef skeleton, much as certain corals do today. Encrusting algae helped to break the force of the waves beating against the reef and filled cavities and crevices with calcium carbonate precipitated in the completion of the algae life processes. Other organisms, such as brachiopods and bryozoans, aided as well.

Slowly the reef grew, until it rimmed the edge of the Permian sea, much as the barrier reefs off Australia rim that continent. Since the coral could not grow out of the water, the low tide level marked the top of its upward growth. In this same period, however, the Delaware Basin was subsiding, so eventually the reef grew to a great height. At times its growth exceeded the rate of subsidence, and during those periods, the reef grew seaward on talus or fragments broken from its seaward face. At the same time, sediments deposited in the lagoon behind the reef were encroaching upon it, burying the landward portions.

At the close of Guadalupian time, when access to the open sea had been cut off by the closing of the Hovey Channel, the reef had advanced seaward from one and one-half to twenty miles from its point of origin. Only a few miles seaward it stood 1,800 feet above the basin floor. The aggregate width of reef limestone and reef talus varied from 1,300 to more than 2,000 feet. It was in this massive structure that the cavities formed which were to become Carlsbad Caverns.

The closing of the Hovey Channel brought the growth of the reef to an end and turned the whole basin into a gigantic evaporative pan in which gypsum, salt, and other minerals were deposited. These filled the basin and the lagoonal area behind the reef, and covered much of the reef as well. At the close of the Permian period further subsidence occurred. The earlier sediments were covered with new deposits later to become a dolomitic cap. Continental deposition and erosion then occurred throughout the Delaware Basin until seas again covered the area, late in the mid-Cretaceous. When these seas receded, there was exposed a low-lying, nearly flat continental area.

Late in the Cretaceous period, at the time of the formation of the predecessors to the Rocky Mountains, stresses and strains resulting from those movements developed the system of joints which was to determine the pattern of Carlsbad Caverns. These joints and fissures formed a network of northeast-southwest, northwest-southeast trending fractures along which ground water movement was channeled. Along these major zones of movement, solution took place at a constantly accelerating pace.

For many millions of years the area remained low-lying and almost flat or peneplained, furnishing the conditions predicated by Dr. W. M. Davis as necessary for the initial or phreatic stage of cavern development. Guided by the system of joints, migrating ground water slowly dissolved away the limestone, eventually forming an intricate maze of small galleries, corridors, and passageways. At an early stage in the caverns' growth, the limestone mass was honeycombed with these solutional cavities.

As solution continued, the partitions between many of the smaller openings disappeared, resulting in fewer but larger passageways and chambers. These cavities were water-filled, and as long as they remained so, the caverns continued to be enlarged. Collapse, too, aided in the enlargement of the rooms and corridors, for as solution went on, walls and floors were so weakened that they frequently fell in, carrying other relatively unstable partitions with them. This collapsing, which played a prominent part in developing the gigantic openings so characteristic of Carlsbad Caverns, was accelerated by the drainage which occurred during the second or vadose cycle which followed.

Late in the Pliocene or early in the Pleistocene, the whole region was elevated. A local movement or fault occurred along the reef, lifting the Guadalupe Mountains and tilting the vast block of limestone containing Carlsbad Caverns gently to the northeast. The mature streams of the old land surface were rejuvenated, etching their way slowly into the limestone cover. A section of the reef, almost forty miles in length, emerged from the softer sediments of the Delaware Basin as a prominent feature of the terrain.

As elevation of the land and erosion lowered the water table, the ground water filling the enormous cavities withdrew, until eventually all of the known galleries, rooms, and corridors comprising Carlsbad Caverns were drained. Collapse continued during and following drainage, destroying all but minor remnants of such cavern levels as may have existed. The extent of such collapse may be judged by the piles of debris which cover the floor of the lower end of the Main Corridor to a depth approaching three hundred feet. Today the trail to the scenic rooms at the 729-foot level negotiates this descent in a series of three gigantic steps—a drop of 160 feet to the level of the Bat Cave, another of approximately 200 to the floor of the Devil's Den, and then down a steep slope, over piles of debris, some 370 feet to the solid floors of the Green Lake Room and King's Palace.

Drainage brought the enlargement of Carlsbad Caverns to an end, except for the collapse of small cavities forming larger ones. It was, of course, this period of collapsing which accounts for the truly tremendous size of the Main Corridor, a gallery a mile in length, forty to more than two hundred feet wide, whose ceilings rise as much as 250 feet above its rubble floors. It accounts, too, for the Big Room,

Giant Dome (left) in the Hall of Giants is the largest stalagmite in the caverns, rising 62 feet. Twin Domes can be seen at right. *Santa Fé Railway*

a huge, T-shaped chamber, whose stem, extending over 2,000 feet from east to west, is capped by a gallery extending some 1,100 feet from north to south. Above the floors the ceilings arch to a height varying from thirty to 285 feet. More than twelve acres of floor space are incorporated under the roof of this one gigantic room.

In its present stage of development, Carlsbad Caverns is in the process of being filled by the secondary structures known as cave formations. It is this wealth of decorative formations which makes it the goal of hundreds of thousands of visitors yearly. Stalactites in an infinite variety of shapes and sizes festoon the ceilings, hanging in huge, threatening masses or dripping like molten wax in long, slender rods. Ribbons of stone decorate the walls and frequently form great, massive draperies, while from the floors a multitude of stalagmites thrust their massive bulks ceilingward. Decorative forms range from the delicate and airy to the monstrous and majestic.

While it is impossible by any known method to date the formations in Carlsbad Caverns, it is believed that none of them are older than early Pleistocene, or about 1,000,000 years. Their greatest development probably took place during wet periods associated with Pleistocene glaciation. Certainly the climate existing when most of the formations were active must have been radically different from the arid conditions prevailing today.

Portions of the cavern, once drained, have since been subjected to recurring periods of submergence and emergence. This is indicated by the partial "decay" or resolution of many stalagmites, by the deposition of clinker-like, calcareous material upon them, and by the enormous deposit of gypsum which once paved the Big Room. Each episode records the return of certain portions of the cavern to submerged, phreatic conditions. For several thousand years, however, the caverns have been as dry as they are today.

Contrary to common supposition, no great stream ever rushed

The entrance to the Hall of Giants is nearly as impressive as the Big Room itself. Many of the stalagmites are covered with clinker-like, calcareous deposits. *Santa Fé Railway*

through Carlsbad Caverns. At various times small streamlets utilized its open passageways on their way to the water table. So feeble were they that, far from having opened these galleries, they were unable even to alter the general, joint-determined pattern. No meander niches have been incised in the wall rock, nor have the stream beds been trenched. These streams entered the cavern long after it had been formed and drained and, at their best, were able to do no more than deposit silt fills and cobble benches along their courses. Water did, however, float in the carcass of a ground sloth, whose bones were found in stream silts below the Devil's Den.

Today, even in the deepest levels penetrated, 1,076 feet below the surface, no running water has been found. As a result, Carlsbad Caverns contains no aquatic life, such as fish or amphibians, for only where drip water has collected in catch basins does any appreciable quantity of water exist, and these pools support no forms of life higher than bacteria and one low form of insect.

For several thousand years after reaching its present condition, Carlsbad Caverns lay hidden in the limestone uplands of the Guadalupes, undiscovered. Then, perhaps a thousand years ago, Indians chanced upon the entrance and gazed into its great mouth. These people, half sedentary, half nomadic, whose culture resembles that of the Basket Makers, made use of the entrance as a shelter and left there a few pictographs on the walls. Their "midden circles," rings of broken, fire-scorched rock, dot the ridge tops in the vicinity of the cavern opening.

More recently, the nomadic Apaches moved into the area, and they, too, knew of the cavern entrance. That these Indians ever ventured into the cavern itself is doubtful, or at least there is no concrete evidence to indicate they did. Early explorers told of finding human bones deep in the cavern, but, in spite of persistent search, none of these have come to light.

Although Spanish explorers moved along the Pecos River late in the 1500's and may possibly have penetrated the Guadalupes in search of precious minerals, there is nothing to indicate that they ever came upon the cavern. Neither did Americans traveling the Butterfield Route, which passed nearby, venture into the forbidding uplands. Following the Civil War, however, settlers began to move into the Pecos Valley, and by the 1870's were homesteading in the foothills of the Guadalupes. By 1885 homesteaders had settled on lands not more than two miles from the cavern entrance, and certainly they must have been aware of its existence.

Perhaps entry was made prior to 1883, for before Rolth Sublett entered it in that year, it was already known locally as "Bat Cave." According to a sworn statement in the National Park files, Rolth Sublett, at the age of twelve, was lowered to the floor of the cavern by his father but did not venture beyond the area illuminated by the sunlight streaming through the entrance.

In 1903 a mineral claim was filed on twenty acres around the entrance to Carlsbad Caverns, then called "Big Cave" by "Bige" Long, a resident of the town of Eddy, now Carlsbad, New Mexico. The object was the removal of a vast quantity of bat guano which had accumulated through thousands of years in a gallery extending almost a half-mile east of the cavern entrance. This was, and still is, the home of millions of bats.

At least two methods of removal were employed. At first guano was taken out on an inclined track through the entrance, but this method proved uneconomical. A short time later, two shafts were cut about a quarter of a mile east of the entrance and the guano removed in ore buckets. These buckets, by the way, served as the first elevators at Carlsbad Caverns. Visitors were lowered in them and from the guano workings made their way precariously over piles of rubble into the scenic portion of the cavern.

Over a period of twenty years, from 1903 to 1923, at least six successive companies were engaged in mining guano. Robert A. Holley of the General Land Office estimated that during that time 100,000 tons were removed. The operation does not seem ever to have been a profitable one, however, and after 1923 removal was sporadic. A mining claim on guano deposits still exists, but no mining has been done in Carlsbad Caverns since 1941.

Jim White, a local cowboy employed by the various mining companies, entered the cavern in 1901. Intrigued by the mystery of the unknown reaches of the cavern, he began to explore it, and over a period of several years succeeded in penetrating most of the now known portions of the cave.

White's descriptions of the cave seem to have met with consider-

"The Clansman" is one of the features of New Cave, one of scores of undeveloped caves within Carlsbad Caverns National Park. *Charles E. Mohr*

able skepticism, but, confident that in time its beauties would be recognized, he constructed crude trails and guided all who wished to come, through the portions he had explored. Skeptics who came to scoff left amazed at what they had seen and singing the praises of the vast, underground wonderland. People in the nearby community of Carlsbad became interested in the cavern as a scenic attraction, and in September, 1922, thirteen prominent citizens made a tour with Jim White. On this trip the first pictures ever to be taken of the cavern were made. These photographs became important in publicizing its beauty.

Within a few months the outstanding natural character of the cavern was called to the attention of the General Land Office, probably by Senator Holm O. Burson or Representative John Morrow, one or both of whom evidently requested a governmental investigation of the area to ascertain the desirability of making it a national monument. In any event, on March 8, 1923, Robert A. Holley was directed to make an investigation of the area. Following a month-long study, Mr. Holley strongly recommended that the cave be preserved as a national monument.

Shortly thereafter Major Richard Burgess, an attorney from El Paso, Texas, toured the cavern. He was so impressed with its magnitude and beauty that he immediately began to exert pressure on various government agencies, members of Congress, and others to make it better known and available for tourists. He was also instrumental in interesting geologist Willis T. Lee in a scientific survey of the cavern.

Finally, on October 25, 1923, President Coolidge signed the proclamation establishing Carlsbad Caverns National Monument and making it a part of the National Park System. In 1930 an act of Congress established it as one of our twenty-eight national parks. A short time later its boundaries were extended to provide protection to other caves in the vicinity, which were being vandalized.

Park Naturalist T. Homer Black silhouetted against the aragonite "tree" formations in Carlsbad Caverns. *Carroll S. Slemaker*

The creation of Carlsbad Caverns National Monument was not, however, tantamount to opening it for wide public usage. Much preparation had to be made before large groups could be accommodated, and little had been done to publicize the cavern as the scenic attraction it was destined to be.

In March, 1924, Dr. Lee, on leave from the United States Geological Survey and sponsored by the National Geographic Society, began a comprehensive exploration of the cave. Almost six months were spent building necessary trails, investigating and photographing its extensive rooms and galleries, and making a thorough survey of the Lower Cave, where more than six miles of passageway were discovered.

Working under difficult conditions and with only lanterns for illumination, Dr. Lee surveyed twenty-three miles of passages and prepared the first map of the caverns. The rooms, galleries, and formations he investigated were given names derived from the Indian mythology of the region; meaningful only to those acquainted with the lore, these names have long since fallen into disuse.

The story of Dr. Lee's first exploration of the caverns appeared in the January, 1924, issue of the *National Geographic Magazine*. This article, followed in September, 1925, by a detailed account of his survey, attracted world-wide attention.

Sporadic exploration by National Park Service personnel has accounted for some new discoveries, but no openings to the surface have been found, other than the original entrance and one small rift near it leading into the portion known as Bat Cave.

One other exploratory expedition deserves mention, not because of its scientific importance, but for its clownish aspects. In 1930 a journalist appeared at the cavern equipped with balloons and rubber boats. How much actual work he did remains a mystery, but his dispatches to a number of newspapers daily reported hazardous adventures, fantastic discoveries, great streams, tremendous abysses, and other wonders which no one else has ever been able to find.

In the spring of 1923, Vernon Bailey, biologist for the United States Biological Survey, conducted a study of life found within the cavern and adjacent to it. He identified ten species of bats, either from specimens or skulls found within the cave, but paid particular attention to the huge colony of Mexican Free-tailed Bats in-

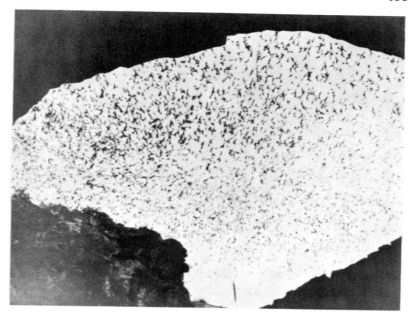

Flight of Mexican Free-tailed Bats emerging from Carlsbad Caverns after flying half a mile through the cave from their roost. *National Park Service*

habiting the Bat Cave. National Park Service naturalists have continued to study the fauna of the cave, and recently an extensive program of bat banding has been undertaken to determine migratory range, intraseasonal migration, and longevity. Bats are caught in nets as they emerge from the cave at dusk, are banded with light aluminum bands, recorded, and released. Although only a very small proportion of the marked bats have been recaptured, one of them represents the record for the longest bat migration ever reported. The bat was taken in November, 1952, at Atengo, Jalisco, Mexico, an airline distance of 810 miles from Carlsbad Caverns.

The bat flights, in many ways, are as spectacular as the cavern itself. For thousands of years the Mexican Free-tailed Bats have made their summer home here. Beginning about mid-April and continuing until mid-October, hundreds of thousands and sometimes millions of bats fly forth at dusk in search of food and water.

Local legend has it that these flights led to the discovery of the cavern entrance. This seems quite possible, for a good flight may last as long as three hours. The bats, pouring from the entrance in incredible numbers, resemble a dense column of smoke as they swing over the escarpment south of the entrance on their way to their feeding grounds. So spectacular is this display that as many as 1,200 people have gathered at the cavern entrance to watch the exodus.

A park naturalist, on duty nightly throughout the flight season, gives a half-hour talk on the bats before the flight begins. The best flights occur, normally, during August and September, when both adults and young are flying. Late in the fall, with the arrival of the first killing frost, the bats desert the cavern for warmer regions to the south, returning the following spring.

Early visitors to Carlsbad Caverns had a long and arduous trip ahead of them. Dropped to the floor of the Bat Cave in guano buckets, they climbed on crude trails over the rubble-strewn floor and down long slopes with only lanterns to light the vast galleries and chambers through which they passed. Even after the cavern became a national monument, trips were made by the light of gasoline lanterns, and it was not until 1927 that the toured section was illuminated with electricity and its true majesty and beauty revealed.

As the use of the cavern increased, a stairway was built descending through the wide arch of the natural entrance. These stairs were later replaced by gentle ramps. In 1931 an elevator was put into operation. Increased now to four, the elevators can evacuate 1,400 people per hour from the cavern lunchroom, 754 feet below the surface.

Although only about three of the twenty-three miles so far explored are open to the public, this portion does include the most majestic and scenic portions of the cavern. The rest is maintained as a research area in which qualified students, interested in various phases of speleology, may conduct scientific studies. The tourist gets a glimpse of one of these areas from the Jumping Off Place in the Big Room, but the unlighted, untrailed sections are deliberately kept as primitive and undisturbed as they were when entered by their first explorers.

The key to all operations at Carlsbad Caverns National Park is

The Christmas Tree, beautiful, six-foot formation in lower-level passage of Carlsbad Caverns. *Carroll S. Slemaker*

caring for vast numbers of people in a restricted area. As many as 7,200 people have entered the cavern in a single day. In few other units of the National Park System do so many visitors use so small a space; hence, the specific methods developed to handle the visitors have had to be geared to the conformation of the cavern and the existing facilities. So integrated is the resulting operation that a single major change may necessitate a multiplicity of minor ones. As an example, the change to private power necessitated a rigid scheduling of pumping operations to keep down peak usage of electrical energy which would increase operational costs. Adoption of the multiple-party system of visiting the cave made it necessary for much of the cavern's maintenance and construction activities to be carried on at night for the sake of efficiency.

At first, while travel was light, visitors were taken into the cavern whenever a sufficient group had assembled, but as travel grew heavier and heavier, it became necessary to set a definite time at which

Tourists emerging from the entrance to Carlsbad Caverns. Visitors now use electric elevators to return to surface. *Santa Fé Railway*

visitors could enter. A system for handling a large party was instituted. Personnel assigned to a party consist of a guide at the front to control the rate of movement and to illuminate the sections to be entered, one at the rear to care for stragglers and to turn out lights in the section left behind, and as many more guides as possible working back and forth in the line to give information and to protect irreplaceable cavern formations. At first only one party a day was scheduled, but as visitors continued to increase, the total rising from 29,034 in 1927 to 284,024 in 1941, it became necessary, particularly on Sundays throughout the summer, to operate several parties. This quickly led to complications, for as parties grew in size and number, difficulties multiplied.

The existing elevators were unable to care for the increased load. Entering parties could pass those walking out on the cavern's single trail only at suitable places, so it became necessary to space party entrance carefully and to adhere to a rigid timetable. When this was impossible, party size was limited, so that one group could be removed by the elevators before another came in to use them.

Permission to take photographs could not be granted on regular tours because of this important time factor. To meet the demand from camera fans, a special photographic trip has been provided. Here, too, the problem of avoiding conflict with other parties on the single trail is paramount. In order to stretch the time during which the scenic portion of the cavern may be seen in its entirety, a short trip through the small rooms at the end of the Main Corridor has been arranged, following an afternoon trip into the Big Room.

The number of visitors continued to increase, rising to 531,831 in 1952—much faster than available manpower to care for them. Additional tours were scheduled, but to reduce party size well below the maximum of 1,800 and to make more efficient use of the personnel available, a ceremony at the huge Rock of Ages stalagmite was abandoned, and the time allowed for completing tours was cut. By such methods it became possible to assign a tour leader to one and one-half trips, instead of the one which had been maximum before this was done. With the installation of new, high-capacity elevators, walkout trips have been abandoned, and with the manpower saved, an additional trip has been added to the regular schedule.

The three-mile tour of Carlsbad Caverns is now completed in three and three-quarter hours, including a forty-minute stop for lunch and rest midway through. At the end of his trip the visitor is whisked by elevator to his waiting automobile.

Thus, the National Park Service spares no effort to see to it that this jewel of caves will be preserved forever in its natural beauty as a part of our national heritage.

The Death of Floyd Collins

In 1925 Coolidge prosperity was beginning to spread throughout the nation. Prohibition had been in force about six years, but bootlegging was rampant. The famous Scopes Trial duel between William Jennings Bryan and Clarence Darrow ended in a defeat for the teaching of evolution. The United States Navy rigid dirigible, the Shenandoah, *was destroyed in a storm with a loss of fourteen lives.*

The Breakers Hotel in Palm Beach, Florida, burned to the ground with a loss of over $7,000,000 in the year's most spectacular fire, and in March a storm in Indiana, Missouri, and Illinois killed 830 people—830 people in states adjacent to Kentucky, where thirty days before, another event had been headlined—FLOYD COLLINS TRAPPED IN CAVE.

For over two weeks the Collins story held the attention of the entire nation perhaps even more intensely, more vividly, than any other of the year.

Roger W. Brucker

In the year 1925 not many Americans could lay claim to the title "cave explorer," but if anyone could, it was William Floyd Collins. Floyd lived on his father's farm, on the Edmonson-Hart County line in the cave region of Kentucky. In his mid-thirties, Floyd had explored more miles of virgin cave passages than most present-day speleologists do in a lifetime, and usually he explored alone, with only a kerosene lantern for company and light. He was a stocky man, muscular, with powerful arms. He spoke little and seldom smiled, but when he spoke, it was with an air of authority that commanded attention.

But Floyd Collins was not a speleologist in any sense of the word; the science of caves little concerned him, except as it might hold a

Floyd Collins, as he appeared shortly before his ill-fated trip into Sand Cave. *Russell T. Neville*

clue for finding a new cave. As a boy, his interest in caves was insatiable, even when the few playmates he had, tired of poking around in cool, damp holes in the ground. Neighbors never forgot the curious sight of a boy tramping over corn fields in the middle of winter, a boy who would lie down on the thin mantle of snow and blow into a "breathing hole," then pause to watch his breath float back out into the chill air. A breathing hole, to most of these country people, was a mysterious thing, a hole in the ground which sometimes exhaled cold air in summer and warm air in winter, and at other times, might suck in air, just as mysteriously. Its only value was as a topic of conversation over the potbellied stove at Sells' Store. To Floyd, however, a breathing hole was a hope and a dream. Beyond that orifice was a cave, and Floyd knew it.

His search for caves led him farther from home, and the words "Floyd's out there a 'pokin' around that old sinkhole," were familiar to households for miles around. When Floyd wasn't looking for caves or exploring them, he talked about them. And when he wasn't talking about them, he dreamed about them, for he loved caves more than anything or anybody. Gradually, his hobby became more than a diversion.

In December of 1917, Floyd began his exploring in earnest in this area where commercial caves—Mammoth, Salts, Colossal, and others —made up one of the region's chief sources of revenue. Veritable cave wars raged in those days. Conniving and cunning were "legal" as long as they produced more paid admissions, but if one cave operator triumphed, he could count the days until retaliation cut into his business, sometimes in the form of a dynamite blast, more often by less drastic means.

When Floyd crawled down into a sinkhole on his father's farm one cold December day, he and his family were drawn into the midst of the struggle, for there he discovered one of the most beautiful caves in the region.

Great Crystal Cave, as he called it, contained three natural attractions which rivaled or excelled any that nearby caves had to offer. First was Grand Canyon Avenue, a large, curving room about eighty feet high, forty feet wide, and nearly three hundred feet long. Second, there was a helictite display of thousands of beautiful stone tendrils in breath-taking colors, a display acknowledged today as one of the most beautiful in the world. Third, Crystal Cave contained gypsum in abundance—gypsum which formed itself into delicate orchid-like flowers, into wall-covering sheets resembling popcorn, and into millions of tiny facets which reflected every light beam.

Floyd and his brothers spent the next several years making trails for tourists to enjoy, but Floyd, not content with what was easily accessible, gradually probed deeper and deeper into the unknown portions of his cave. On these exploratory trips Floyd usually traveled alone. Invariably, his brothers were too exhausted after a hard day of trail-building or working on the farm to want to go with him. So, Floyd would fill up the tank of his coal-oil lantern, jam a can of baked beans in his pocket, and disappear into the dark recesses of the cave. In time he became used to exploring alone; in fact, he came to prefer it. He certainly knew the risks, for he had several brushes with death. Once, when he jumped off a ledge into a pit, he found that he could not climb back out. Resourcefully, he piled up rocks, probably a job of several hours, until by standing on the pile he could pull himself up. On another occasion, he ran out of kerosene deep within the cave. Because of his familiarity with the labyrinthian passages, he groped his way to daylight in the space of thirty-six hours. The same trip, with light, takes only an hour. To avoid these inevitable "solitary excitements," today's explorers travel with company and extra light.

In March, 1922, Floyd returned from one of his prolonged explorations to announce that he had discovered a large passage a mile long, as big as a railroad tunnel in some places. The news of Floyd's discovery startled no one, for he had already gained a reputation among his neighbors as a kind of superman when it came to cave exploring. Over and over they told the story of how he was wedged in a passage for twenty hours before his brothers managed to free him. After that, most men would have foresworn caves, but not Floyd Collins. He loved cave exploring above security and comfort.

Despite the beauty of Crystal Cave, comparatively few people came to see it, primarily because of its remoteness and the miserable two ruts leading to it, which were all that could be called a "road." To be successful, a commercial cave should be located on a main road. Since the Collins family owned no property on the main road, Floyd sought a different solution to the problem.

After thirty years he knew where every ridge ran, where the solutional valleys lay, and by talking to the geologists who visited the cave, he gained some practical understanding of the peculiarities of limestone geology. Crystal Cave was located under an arm of Flint Ridge, the largest cave ridge in the area. Immediately west of Flint Ridge is Mammoth Cave Ridge, and five miles south of the Collins' home the two ridges join. It is significant to note that they join at the main road at the highest point of both ridge systems. The discovery of a cave here could mean the combination of two vast cave systems into one, and consequently the discovery of the largest known cave system in the world. What a discovery that would be! Cave passages heading northeast would connect with Crystal Cave. Any passages head-

The entrance to Sand Cave as it appeared in 1925 after Collins' body had been removed. *Russell T. Neville*

ing northwest could only connect with Mammoth Cave. Collins knew these facts well and acted upon them.

In the first week of January, 1925, Floyd Collins signed a contract with three of his neighbors, Bee Doyle, Edward Estes, and J. L. Cox, all of whom owned land over the junction of the two ridges. Under terms of the contract, Floyd was to begin the exploration of a cave located on Doyle's property on the southern slope of the ridge junction. Floyd told the three farmers that he believed the cave extended under their land, and that for a half-interest in the cave, should it be there, he would explore it for them. The farmers agreed, for commercial caves in the area had made others rich, and in Floyd they knew they had the best explorer around. For three weeks Floyd had "worked" the cave, an ugly hole leading into blackness under a sandstone ledge, which he called Sand Cave.

On a gray Friday morning, January 30, 1925, Floyd ate breakfast at Doyle's house. His good friend Doyle listened as Collins reported on his progress. The previous day, Floyd had set off a dynamite blast to enlarge a crawlway in Sand Cave, and today he thought he would find what he had been looking for. Doyle suppressed his excitement with difficulty, as Collins shouldered his sack of tools and left the house.

Under the ledge at Sand Cave, Floyd lit his kerosene lantern and put on a denim jacket. He crawled for about thirty feet, then flattened himself on the mud floor and continued. He rounded the "squeeze," a narrow S-turn, which brought him to the scene of the dynamite blast. With a short crowbar, he pushed the debris aside, then paused to assess one particular piece of breakdown, about as big as a large watermelon. The boulder was pressed against the ceiling by a wedge of loose rock, allowing just enough room to squirm beneath it. Cautiously, he wormed his way under it, pulling his crowbar after him. Now he was at the edge of a pit leading to the blackness of a bigger passage on a lower level. He climbed down, using the rope he had placed there earlier as a handline.

At the bottom he continued, this time walking in the same direction in which he had been going. Soon he turned into a left-hand passage, and there he found a "big cave." Exactly what he found may never be known, but those who talked with Collins afterward repeated his description of "gleaming gypsum" and "stalactites." At any rate, on this unfortunate Friday morning Floyd found the large cave he knew was there, and because of his blasting to enlarge the crawlway, he felt that he could now bring his three sponsors to see this underground wonder. Quickly, he started back to get them. Up the pitch he went, hanging onto the handline. He started into the crawlway leading to the outside world. And then it happened.

In the narrow crawlway his foot kicked and dislodged the wedge rock. There was no awful roar or cataclysmic tremor; there was probably just a hollow "clunk" as the melon-shaped rock slipped a few inches toward the floor, coming to rest in part on a vertical ledge of rock and in part on Collins' left ankle. At first, this obstruction was no more than a nuisance; it was not even particularly uncomfortable. Yet, try as he would to move his powerful leg, the rock could not be budged. He remembered his crowbar! "Chink, chink, chink," he tapped ineffectually at the rock. It would roll back and forth, but the cursed crawlway was so tight he could not work himself into position to make good use of leverage.

For an hour he strained and struggled with this nuisance of a rock which pinned him down in the damp, uncomfortable crawlway. The

Man in actual hole where Collins was trapped. *Russell T. Neville*

result of his efforts was only to dislodge more rocks from the shattered ceiling. He saw that he could not free himself without help, and that further attempts on his part would only make matters worse. He sloshed his lantern, mentally calculating that his light would last until the middle of the afternoon. There was nothing to do now but wait for the help he knew would ultimately come.

That night, the clouds burst and lightning split the dark sky. Rivers of water poured down gullies and into crevices. Sixty feet beneath the surface of the earth, Collins tried to avoid the steady drip of water falling on his face, but he was pinned tight and could not move. When would his friends come? Perhaps not for days, for by now they were used to the long absences of this solitary explorer. Many times he had chided them for worrying. Now he wished he hadn't.

When Bee Doyle failed to find Collins in bed the next morning, he went to Estes' home. Not finding him there either, Doyle, Estes, and Estes' son, Jewell, trudged through the wet woods to Sand Cave. Reluctantly they crept into the cavern. Jewell, aged seventeen, went ahead because he was "smaller," and just beyond the squeeze, the lad called out.

Collins answered, "Go back out and hurry over to the home-place and tell the folks. My legs are caught by at least one rock, but if you have my brothers come over here with some of the other boys, I believe I can get out of here without too much trouble."

On horseback, Doyle and Estes reached the Collins' home Saturday noon. Lee Collins, Floyd's father, was furious. "You are to blame," he told them; but then he collected himself. Floyd's brother, Marshall, organized a rescue party, while others went to the nearest telephone to call Homer Collins in Louisville. Homer was closest of all the brothers to Floyd and was considered the second-best cave explorer in the family.

When Marshall's rescue party reached the cave, they found a semicircle of curious farmers in front of the entrance. All afternoon, the party-line telephones had been jingling with the news, and now, the group around Sand Cave was beginning to resemble a gathering of the clans. Armed with a long crowbar, Marshall pushed into the cave, followed by several others. For three hours they pried and worked to extricate Floyd, but the passage was too cramped, and what little space remained was blocked by his body. They did manage to divert the dripping water with the aid of a rubber raincoat, thus ending the period of "Chinese water torture," but another more horrible torture followed.

As the sun set that Saturday evening, Homer Collins arrived at Sand Cave in a battered Model T. Confidently, he organized a new party to travel the tight, muddy crawlway. While he toiled to release his brother, his companions retreated to the entrance. Homer, in eight hours, removed several bushels of loose rock from around Floyd's chest and waist, and by his encouraging words bolstered Floyd's belief that he would soon be freed. But eight hours of rock-moving in a cave, coming at the end of twenty-four sleepless hours, took their toll on Homer, solidly built as he was. Exhausted, but reluctant, he crept out of the silent hole into a world of confusion and chaos.

That same Sunday morning, a farmer entered Sand Cave, then emerged to tell the crowd that he had felt "the cold sweat of death" on Floyd's forehead. A man from Tennessee then went in and came back to pronounce the report false. He had given Floyd some hot coffee and had carried on a conversation with him. Sporadically, parties entered the cave, but always came back to report no progress in freeing the trapped man. By Sunday evening, moonshine was flowing freely through the veins of the would-be rescuers, and the spectators ringing the mouth of the cave became a reeling, brawling mob.

On Monday morning, newspaper reporters began to arrive. First came a man from the Louisville *Herald*. He entered the crawlway and carried on a conversation with Floyd that became the meat of the first eyewitness account. He was followed closely by cub reporter William Burke ("Skeets") Miller of the Louisville *Courier-Journal*. Then came the first of the "experts," Lieutenant Robert Burdon and Private John Blake of the Louisville Fire Department. Burdon believed that the only way to rescue Collins was to sink a shaft to him from the surface, but no one paid attention, for the drunken crowd was getting out of hand.

Inside the cave, Floyd Collins rested well that Monday night. He had eaten three meals in the course of the day, though they consisted only of liquids, and was feeling better about his plight. The world was coming to his rescue, spurred on by dramatic newspaper accounts of the epic struggle—at least, that was the word brought to Floyd.

Late Monday night, Burdon conceived a bold plan. Homer Collins fashioned a leather harness which he, Burdon, and Miller dragged into the damp cave. They fitted the harness around the upper part of Floyd's body while they explained to him what they proposed to do. By pulling on the rope attached to the harness, they hoped to drag Floyd from his prison.

"Go ahead and pull," said Collins. "I'd rather be dead than down here."

Lying prone, the three men heaved desperately at the rope. Floyd groaned in anguish, but his body didn't move. Miller and Burdon continued to pull, but Homer slacked off, unnerved by the moans.

"Don't do it," cried Floyd, "I'd rather die here than have my foot pulled off."

Thwarted, the trio left the cave. Outside, Homer offered $500 to any surgeon in the world who would amputate Floyd's foot, but the offer had no takers.

In any chaotic situation, a leader is likely to emerge if the confusion continues long enough. Into the leadership vacuum stepped John Gerald, a friend of Lee Collins' and a cave explorer of some ability. Gradually, he began to direct the comings and goings of the curiosity seekers, the "experts," and the rescue parties. Inevitably, such a leader, under stress, must make quick decisions. When two monument-makers arrived to help, loaded with chisels and hammers, he refused to allow them to enter, probably because he believed they were no different from the host of other "experts" who had volunteered assistance, entered the cave, and left, their curiosity satisfied. Or perhaps Gerald felt that the insecure-looking ceiling would certainly not hold up under steady pounding. At any rate, the monument-makers left in a huff, and the ineffectual rescue attempts continued.

Two strong men accompanied John Gerald into Sand Cave to try once again to free Collins. Working for seven hours, they removed enough rock to permit a small man to creep down beside Floyd with a jack, in an attempt to move the key rock which held his foot. Several firms had sent jacks of all sizes to the site—everything from giant, house-moving jacks to tiny models, barely bigger than a hand.

Miller and Burdon led in, both clutching jacks in their hands as they squirmed over the cold, slippery floor. "Skeets" Miller, smallest of all the rescuers present, selected one jack, then ducked under the ledge, spreading himself along the length of Floyd's body. Scraping aside the rubble, he was able to see the watermelon-shaped key rock. Slowly, he tried to slip one of the jacks into position. It was too big. Collins urged him to try another. The smallest of the jacks was passed forward, and again Miller moved it along the length of his body, past his head, and into the small, black opening ahead of him. With a tiny wrench, he began lifting the jack.

The rock was going up! The three men were breathless as Miller's hand moved a fraction of an inch with each turn. Then the jack slipped, and the rock settled back.

Again Miller tried to brace the jack. Again it slipped. He continued struggling, but each time the jack would slip on the rounded side of the rock. It was like trying to jack up a globule of mercury. This struggle, in terribly cramped quarters, continued until Miller was physically and mentally exhausted. How much easier it would have been, had the jack not been able to move the rock; move it, it did, but raise it, it would not. At last the two men turned back, leaving behind them a frightened man pleading for companionship in a remote crevice in Sand Cave.

On Tuesday afternoon, February 3rd, another expert arrived,

noisily heralded by the press. Henry St. George Tucker Charmichael was an engineer with the Kentucky Rock and Asphalt Company, and besides bringing to the problem his own technical know-how, he led a hand-picked crew of workmen. Immediately his men began timbering the ceiling of the cave from the entrance in, preparing to bring Collins out that way.

On Wednesday, "Skeets" Miller started in on his seventh trip, carrying milk to Floyd, and a cable with an electric light to provide some warmth for his now shivering body. John Gerald went in after Miller returned, and emerged to announce to the press that a section of the ceiling between the squeeze and Floyd had collapsed. The impassable block prevented anyone from reaching Collins with food.

Gerald repeated Floyd's words: "I'm tired, Johnny, so tired, and I'm going home and go to bed."

Charmichael went into action. For four days no one had thought to make a survey to the spot where Collins was trapped. Now Charmichael directed a survey team into the cave to get data that would enable them to sink a shaft from the surface to where Floyd lay. Roy Hyde, leader of the team, pushed the survey to the edge of the rockfall, about fifteen feet from Collins. Hyde called out to Floyd and heard him moan, "My God, why don't you take me! Mother— I am coming home in just a little while."

"Hold on, old man, we're coming," shouted Hyde.

"You're too slow . . . too slow."

Those were the last words anyone ever heard Floyd Collins speak.

By Thursday the area around Sand Cave was a maelstrom of human activity. National Guard troops had arrived that morning to keep order, and though the bootleggers faded from the scene, the throng of curious spectators had grown to the hundreds. Tents housed some of the group, and others stood in knots in the intermittent rain. Lunch wagons fed them, at exorbitant prices.

Rumors spread like brush-fire. One claimed the whole thing was a hoax, and that Floyd left the cave every night. This rumor was reported as fact by one wire service. Other rumors persisted that Floyd was dead long ago, or that he had just been rescued.

But in spite of this, Henry Charmichael started his shaft at one-thirty Thursday afternoon, February 5, 1925. Day after day, the digging and timbering continued, the work crew being augmented by volunteer labor from the surrounding countryside. A hoist car-

After Collins' death this hut was erected near the shaft from which his body was extricated. *Russell T. Neville*

ried up the rubble; then a rail car hauled it away to a dump on the edge of the ridge. It was slow work, and the nation which had been so tantalized by the newspaper reports earlier in the week wondered why fewer reports were now forthcoming. Was word being withheld? Would they reach Collins? Had they already found him?

On Sunday, February 8th, a caravan of vehicles converged on Sand Cave. By noon, 10,000 people crowded the area, and by three o'clock the number of sightseers had reached 15,000. The line of parked vehicles extended for four miles in each direction along the main road, and thousands of other cars were parked in nearby fields. "Sand Cave" balloons were hawked by vendors, and the lunch wagons ran out of food.

On Monday morning, a Military Board of Inquiry met to pin down and separate fact from fiction. They blamed the fiasco up to this point on excessive drinking, inept and chaotic rescue attempts, and haphazard organization.

Another week went by, in which the shaft approached the sixty-five foot level, at a spot where Charmichael predicted they would find Floyd Collins. The crowd had thinned now. Minor crises in the progress of the shaft were overcome when collapsing timbers were speedily re-braced, and canvas over the excavation diverted rain from pouring into the pit.

At one-thirty, Monday afternoon, February 16th, one of the workers made the breakthrough to the cave from the bottom of the shaft. With another companion, he lowered himself through the hole, then crept past the rope to find the lifeless body of Floyd Collins, lying where the rock had trapped him.

John Gerald went down to identify the man. The coroner ruled that Floyd had died between two and four days prior to the completion of the shaft. Friday, February 13th, fell within this period, and this was the last day on which anyone in the crawlway below had claimed to have heard the sound of Collins' breathing. The doctor attributed the cause of death to "exposure and exhaustion."

While consideration was being given to the disposition of the body, one newspaper photographer handed his camera to a rescue worker, with instructions on how to make a last photograph of Floyd. The worker pressed the button as directed, and the plate was delivered to an aviator standing by to fly it to a Chicago paper. Charles A. Lindbergh was the aviator, but unknown to him, someone switched plates. At the Chicago newspaper office, the editors gazed dejectedly at a blank film. Such was the cutthroat competition in the scramble for the last Floyd Collins scoop. Later "Skeets" Miller received the Pulitzer Prize for his eyewitness journalism, although his heroic efforts to free Collins were even more outstanding than his journalism.

By family agreement, the shaft was sealed with Floyd Collins' body still resting in Sand Cave. But in April, they expressed a desire to give Floyd a proper burial, so the shaft was reopened and the body removed to a hill overlooking Floyd's beloved Crystal Cave. Later the body was moved to a bronze casket in the center of Grand Canyon Avenue in Crystal Cave, where it rests today.

In retrospect, it is obvious that Floyd Collins violated every rule observed by experienced cave explorers today; he explored alone, he carried no spare sources of light, and he seldom told others where he was going. It is equally obvious that no one was prepared or able to institute adequate rescue attempts. Today, there are trained, well-equipped rescue teams, able to reach any part of the country within hours. The debacle of Sand Cave need never be repeated.

And what of Floyd's discovery? To this day, no one has been

Collins' tombstone, before his body was removed to Floyd Collins Crystal Cave. *Russell T. Neville*

back to explore the cave. Its entrance is effectively blocked. Other approaches may be possible, and one day, perhaps, the headlamp of an explorer will fall upon a rotted rope hanging down from the top of a pit where Floyd rigged it. A few feet farther on, he may come to the gallery that Floyd found, and in so doing, may link two great cave systems into one. With prudence and care, the explorer need not be victim of a watermelon-shaped rock and well-meaning but inept rescue attempts. This is the legacy that Floyd Collins left to speleology.

Bones and a Railway

The unusual circumstances which surrounded the discoveries made in Cumberland Bone Cave, Maryland, could hardly have been invented by the most imaginative fiction writer. A Catholic Brother with the appropriate scientific training, a cooperative railroad on whose property the cave was located, and, most important of all, the fact that the cave contains valuable paleontological specimens combine to create a story of equal interest to scientist and layman.

Brother G. Nicholas belongs to the Christian Brothers, an order devoted to education. Brother Nicholas holds his B.S. and M.S. degrees and is completing work on a Ph.D. He teaches religion, physics, chemistry, and biology at LaSalle High School in Cumberland, Maryland. Though only twenty-seven years old, he is Vice-President for Research for the National Speleological Society and Associate Editor of the American Biology Teacher, *a member of Phi Beta Kappa and Sigma Xi, as well as half a dozen scientific societies. He has written numerous papers on botany, biochemistry, education, paleontology, and speleology, and has appeared on many radio and television programs. In his spare time he works on Cumberland Bone Cave.*

Brother G. Nicholas, F. S. C.

The Cumberland Bone Cave would still hold its secret were it not for a railway. To explore this cave it was necessary to abide by regulations found not only in the speleologist's manual but also in a book of rules published by the Western Maryland Railway. What at first appeared a hindrance eventually proved a boon, for the final stage of discovery would have been impossible without the assistance of the engineering and operating personnel of the railroad.

For an understanding of these unique conditions, we must go back some 20,000 years and stand, in imagination, on a weathered limestone ridge overlooking a wide river valley in western Mary-

172

land. Below this ridge, today, lies the small town of Corrigansville, northwest of Cumberland, and the wide prehistoric river has dwindled to a puny stream known as Wills Creek.

On our imaginative trip, however, we are viewing a land of vast evergreen forests and mighty waterways. The melting foot of the last of the great ice sheets lies but a few hundred miles to the north of us, and the glacial outpouring has filled broad river valleys. Centuries later, the flow will greatly diminish as the ice retreats hundreds of miles northward and the waters find shorter routes to the sea. The climate in 20,000 B.C. is similar to that of northern Canada today.

A modern geologist might inform us that this limestone ridge is part of the deeply dipping west flank of the Wills Mountain anticline and that the rocks themselves are Devonian, 313,000,000 years old. As we walk along the northern spine of this ridge on the western side of Wills Creek Valley, a geographer might note that we are traveling on what was to be known as Andy's Ridge, a small hill worn down to a maximum elevation of 1,300 feet and sandwiched between two higher mountains: Wills Mountain on the east and Little Savage Mountain on the west. A biologist might observe that at a slightly earlier epoch this valley was one of the main pathways to the Potomac Valley for animals migrating southward to avoid the cold from the advancing ice sheet—a gateway through the Appalachians to the warmer southern plains. Later these northern animals were to reverse their path.

Continuing our journey down the ridge to the river, we see many animals clustered at the water's edge, drinking at a quiet eddy. Muskrats and otters frolic along the banks, while at the edge of the dense woods a small herd of deer moves cautiously, fearful of the pumas which occasionally creep down from the surrounding mountains. Looming over all is the bulk of a mighty mastodon, the ancestral grandparent of today's elephant, carefully herding several smaller ones into the water.

On one of the many tributaries, beaver are industriously blocking the stream with a dam made of branches and trunks from the birches and maples growing on the sandy river banks. In the treetops squirrels scurry about, ignoring the porcupines and weasels crouching on the lower limbs. The floor of the forest is tunneled with the runways

It is this type of entrance into which many prehistoric creatures fell. Their bones gradually washed deep into the recesses of the cave where modern speleologists may find them. *Charles E. Mohr*

of shrews, mice, and chipmunks, which are routed by an occasional skunk or fox. Among the duckweed and water lilies near the shore are frogs, tadpoles, and other forms of aquatic life which often serve as food for the prowling mink and weasels. Families of bear, and bands of elk, moose, and wood bison tramp through the valley gateway.

Looking back up the ridge, we note that the rocks have been pushed up to form a path from the top of the ridge down to the water hole. Near the edge of this path is a sinkhole, partially obscured by the heavy forest growth. This sinkhole opens into an irregular shaft over ten feet wide which drops seventy feet into a large cave. It is a rather irregular chimney, rising from a huge room.

The floor of the cave is covered with fragmented skeletons and bones. The sides of the sinkhole are draped with decaying carcasses that gradually fall to the bottom to join other layers of floor coverings.

The remains of all sorts of animals, from mastodons to bats, are entangled there, the accumulation of thousands of years. Indeed, so long a time has elapsed since the first creature fell into this natural trap that even the climate has changed completely.

Beneath several feet of mud, washed in from the surface, are fossils of such animals as tapirs, antelopes, and alligators—relics of a time when the climate was much warmer and the area was semitropical. Through the mass of rotting flesh and bone, cave rats and deer mice crawled, cleaning off the flesh, often leaving tooth marks on the bones.

With the end of the ice age the sun grew hotter with each succeeding century, the ice sheet gradually retreated northward, and the heavily furred animals of the arctic regions, mastodons, moose, grizzly bears, and others, followed the retreating glaciers. The river which had been cutting the valley deeper for dozens of centuries, gradually carried less water. The sinkhole, originally close to the water, was left by receding waters on a bluff some two hundred feet above the diminishing stream and a quarter of a mile away from it.

The leveling process of erosion eventually filled the sinkhole almost to the surface with mud and gravel. The path-like channel which led originally to the waterhole now, in this later age, leads to a hillside covered with well-rounded boulders and gravels of quartzite and conglomerate. The graveyard of animals, having silently sealed itself shut, protecting the bones within, kept its secret until disturbed at last by the coming of man—and the railroad.

We have journeyed back 20,000 years to visualize the beginning of the Cumberland Bone Cave. Now we must jump to 1910, when engineering crews of the Western Maryland Railway, surveying for a route westward from Cumberland into Pennsylvania, noticed that the Wills Creek Valley provided a passageway through the Allegheny Mountains, just as it had thousands of years ago for animals alternately fleeing the numbing ice or following the warming sun northward. A practical route with reasonable inclines had been plotted, but a single limestone ridge projecting into the valley blocked the logical path of the rails. In fact, the surveyor's line led right over the sinkhole. If this ridge could be bisected, a route could be laid that would provide the shortest way through the mountains.

Almost immediately, specifications were drawn and blasting begun for a seventy-foot cut, several hundred feet long, through Andy's Ridge. Had a decision been reached to place the new line someplace else, the Cumberland Bone Cave would probably still hold its secret.

As the excavation of the railroad cut proceeded, the cave beneath the sinkhole was being cut in half. Dynamite blasted open the layers of rock and the fill of mud and bones. How many bones were destroyed or carried away in the rubble excavated from the cut no one will ever know. As the cut was almost down to the level of the main room of the cave, an amateur naturalist, Raymond Armbruster of Cumberland, was watching the operations. Suddenly he noticed several huge bones being removed by a power shovel. He quickly notified J. W. Gidley, the vertebrate paleontologist at the United States National Museum in Washington.

Gidley hurried to the scene and immediately began the tedious work of removing what bones remained. His job continued intermittently over four years, and he carted back to the National Museum one of the largest collections of Pleistocene mammals ever found in the eastern United States. Pleistocene remains are 600,000 to 1,000,000 years old.

After the railroad bed was established, thus ending his downward probing, Gidley dug into the part of the cave that had been exposed on the south side of the cut. Apparently he never noticed evidence of a cave on the north side.

Gidley was forced to stop work in 1916 for fear that the whole cave wall might collapse. The bones he found were so imbedded in the lime concretions that large quantities of rock had to be removed to obtain them. In doing so, some of the supporting columns and walls were undermined, making the cave hazardous for inquisitive sightseers. The narrowness of the cut prevented people clustered near the entrance from getting a safe distance away from passing trains. This, and the unstable condition of the cave itself, led the railroad to close the cave by dynamiting the entrance shut.

The remaining fossils might have lain undisturbed for perhaps another 20,000 years except for an apparently unrelated discovery—and, again, the railroad. In 1948, while exploring a cave near Harper's Ferry, West Virginia, I found the femur or leg bone of a bison imbedded in the mud. The only other such fossil recorded in the area was one reported by Gidley from the Cumberland Bone Cave.

The original entrance to Cumberland Bone Cave, Maryland *(upper opening)*, and the enlarged secondary entrance on the near side of the tracks. *William P. Price*

So in November of 1950, I made my first trip to the Bone Cave, discovering its location with little difficulty. But the blast of dynamite had effectively sealed the cave. Even the help of half a dozen of my students from LaSalle High School was insufficient to re-open it. The pile of big boulders in the first passage was too formidable an obstacle. I disliked giving up, for there was good reason to believe that there were still important deposits untouched.

Studying the formations now exposed to the weather on the previously excavated south side of the cut, I noted that the slope was such that if the cave did continue in the opposite direction, across the tracks and beyond, the passageway should be thirty feet above the roadbed. Following my hunch, I found remnants of stalactites and flowstone—proof that the cave actually had extended in this direction across the tracks. Climbing back to the original bone cave and sighting along an imaginary line extended across the cut, I spotted a small opening under a ledge of rocks.

Excited now, I climbed back up this northern side of the cut. I was able to reach the opening, but it was much too narrow for me to squeeze through. Perhaps, I thought, it was merely a natural crevice due to erosion. Almost absent-mindedly, I reached into the crack. My hand groped over the dried mud typical of caves. Suddenly I felt something hard, grasped it, pulled it out into the light. I couldn't believe what I saw—a perfectly preserved left half of the jaw of a giant grizzly bear. Surprisingly, its teeth were intact. This was all the proof I needed that this was indeed a continuation of the bone cave.

Months later I learned of an amazing coincidence. Six months earlier Theodore Ruhoff, an amateur archaeologist from Alexandria, Virginia, had climbed to this very same opening, reached into it and pulled out the jawbone of a bear. Comparison of the two jaws showed that they were not parts of the same bear; in fact they were completely different species! The first jaw belonged to a black bear, mine to a grizzly! Since Ruhoff was engaged in excavating another bone cave in central Virginia, he was unable to investigate the possibilities at Cumberland Bone Cave; however, he very generously donated the bear remains, and other bones found nearby, to the collection that I was amassing from the site.

Every weekend during the winter and spring of 1950-51 students and friends worked with me at the "new" bone cave, gradually excavating a remarkable series of fossils. After widening the entrance so we could squirm inside, we formed a room twenty feet long, ten feet wide, and two feet high. Careful digging in the dirt floor soon exposed additional bones.

An immediate problem arose; how could we dispose of the dirt? Space was so limited that shoveling dirt and rock was an extremely arduous task, and a dangerous one, I discovered.

One day, while I was balancing a mass of rock some two feet in diameter it rolled over and wedged itself in the entrance so solidly that neither I nor two others with me in the cave could budge it. Fortunately, I had always insisted that one member of our party remain outside on call in case of such an eventuality. Our emergency man quickly came to our rescue. Earlier, we had noticed another small opening six feet above the present entrance. From inside the cave it seemed that this crevice was blocked by only a few rocks. Our companion on the outside started pulling these out and soon exposed a chimney-like passage. It proved just large enough for us to climb out.

Through our combined efforts, we finally succeeded in dislodging the boulder that blocked the original entrance. Now, since we had two entrances, we shifted mud and rocks toward the first, and entered through the other at the top of the chamber.

By the summer of 1951, parts of the room had been excavated to a depth of two feet, but work progressed slowly. Each shovelful of dirt had to be sifted because remains such as bat jaws and shrew ribs are only one-eighth to one-fourth of an inch in length and can easily be overlooked. Once below the first foot of fill, we began to unearth bones with encouraging frequency. Some of them obviously were of recent origin, but many others were heavy with mineral impregnation.

By April, 1952, we had excavated so much rock that we suspected we might be undermining the whole cliff. Because of the mountain grade, the long freight trains had two, three, or sometimes four locomotives. Every time a train rumbled by, the whole cave would shake as in an earthquake. In addition, ventilation was so poor that even with respirators our work was seriously handicapped. But if the upper strata of rock could be removed down to the cave level, the danger of collapse would be eliminated and the problem of ventilation would be solved. All we needed was a steam shovel, a crew of men, and a train to cart away debris!

Erosion would eventually weaken the top layer of rock and endanger the railroad, I argued, in talking with members of the engineering staff of the Western Maryland Railway. A slide in this cut could prove disastrous, I explained, since it might go unnoticed until a train was too close to stop. This, they admitted. The cut curved to each end, giving only a short view ahead. This and other considerations, particularly the interest shown by George Haworth, vice president and general manager, led the Western Maryland Railway to agree to excavate the rock down to the level of the cave floor.

In November, 1952, a work crew, consisting of six laborers and a foreman, was assigned to the biggest cave excavation job ever undertaken in the eastern United States. A work train arrived with an air compressor, several hundred feet of pipe, pneumatic drills, cases of dynamite, rope, hammers, chisels, and other equipment. Camp cars for the workmen were placed on a siding several miles down the line from the cave. A telegrapher was assigned to the crew, and connections made to the wires running alongside the tracks. This was an obvious safety measure, for the exact time of train movements had to be transmitted to the work crew so that they could take shelter and avoid blasting when a train was approaching. A severe snow storm and cold weather hampered operations for several weeks. Progress was depressingly slow, since only small charges of dynamite could be used. Otherwise, bones lodged in crevices would have been pulverized. The work pattern that finally evolved was to drill some ten feet into rock, pack dynamite in the hole, check to see if trains were due, set off a charge that merely loosened the rock, and then pry apart the fractured rock with crowbars. Rock which fell on the tracks was piled alongside the track, against the sides of the cut. When a great quantity had accumulated, a work train would be dispatched and a crane would load the rock into gondolas. These were hauled five miles up the track and the rubble dumped on an unused siding.

No less than 2,000 cubic feet of rock had to be excavated to reach the floor of the cave. Then the walls were removed, a little at a time. Today the site bears little resemblance to a cave. There is a cut forty feet deep, twenty feet wide, and about sixty feet long. At the bottom lies a great quantity of clay fill which we have not yet investigated.

Early in 1953, paleontologist LeRoy Kay, of the Carnegie Museum, Pittsburgh, visited the site and invited me to use the facilities of the museum for the necessary cleaning and identification. The thousands of bones we had unearthed since 1950 were mainly fragmented, intermingled with clay and breccia.

The cleaning process was tedious but not difficult. After the dirt had been scrubbed off, the bones were placed in dilute hydrochloric acid. This dissolved the calcium deposits around them but did not affect the bones because the silica which had replaced the mineral matter within them was impervious to the acid. Many of the bones had been so gnawed by rodents that identification was difficult or impossible. All told, however, forty-five different species of mammals have been recorded. Of these, twenty are extinct—including a mastodon, a new species of big-eared bat, and a new type of pocket gopher. Many of the others no longer inhabit the region but may be found in other parts of the world—peccary, puma, tapir,

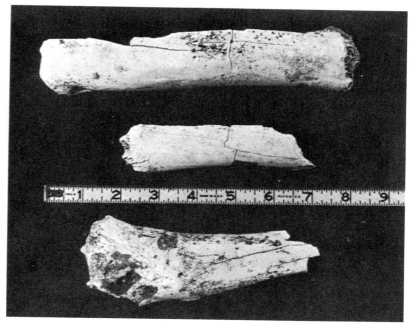

Ancient peccary bone fragments found in Cumberland Bone Cave, Maryland. *William P. Price*

and elk. Besides these mammals, a type of ruffed grouse has been identified, and reptiles include snakes, turtles, and crocodiles.

The description of animal life of the Ice Ages given in the first part of this account is based entirely on remains found in the Cumberland Bone Cave; however, this reconstructive scene has been corroborated from several lesser cave collections.

Bones of paleontological significance were discovered not long ago in Jones Quarry Cave, situated near the town of Falling Waters, West Virginia, about ten miles north of Martinsburg. Although this cave had been known for many years and a portion of it had been surveyed, it was not until November of 1954 that a speleologist uncovered the passage that led back to a crypt containing a jumble of bones. These were tentatively identified as Indian bones some four hundred to five hundred years old, but closer examination suggested that they were between 2,000 and 3,000 years of age.

As soon as the new dating was completed, the Pittsburgh Chapter of the National Speleological Society, under the direction of Richard Hoffmaster and aided by a group of scientists from the Carnegie Museum, undertook plans to remove the remaining bones. Miles Thompson, Jr., an amateur naturalist, and I joined the group of approximately two dozen scientists at Falling Waters in early March, 1955.

Previous reconnaissance of the cave by members of the Pittsburgh Chapter showed that it would be too difficult to carry the bones all the way back to the entrance. Two crawlways some eighteen inches in height (one of which ended in a drop of fifteen feet), and the abundance of mud, would make the removal awkward. It was noted, however, that a sinkhole on the surface apparently lay directly over the bone crypt. A telephone line was stretched from this room to the entrance of the cave and back over the sinkhole. This permitted persons in the cave to report the locations of the sounds of digging in relation to the cave room.

I was immediately impressed with the effectiveness of this orientation system. We had hardly reported "you're right over us" when a well-directed blow from above dislodged a rock fragment right over us. Fortunately, as is customary, I was wearing a hard safety hat, so I suffered no injury. I could testify that those on the surface were aiming in the right direction.

It was decided, finally, to use dynamite to penetrate the layers of rock between the crypt and the sinkhole. A tarpaulin was carried in, and two of us placed it over the bones so as to protect them from any damage.

The only apparent result of the first blast was blowing apart a bucket accidentally left near the hole. With the second shot, however, a column of smoke swirled into the air, indicating that a draft of air from the cave was escaping to the outside. After a few minutes spent in clearing away the jumbled rock, we were able to enter easily.

Almost immediately we began to find Indian bones. Several large femurs or thigh bones obviously belonged to full-grown adults. The first skull, partly intact, was judged to belong to an Indian girl about fourteen years of age. A number of other skull fragments were discovered, but no complete skulls. The long period of time and the less than ideal conditions for preservation had caused disintegration of the lighter skull bones. The bodies of the Indians were probably brought into the cave at a time when the sinkhole was still open. We discovered them on shelves or on the floor of the cave, but not in positions such as to suggest ceremonial burial. This and the fact that no complete skeletons were found, might, of course, mean that the bones were disarranged by animals that entered the cave.

Notable among the artifacts found scattered with the bones were a few breast pendants usually worn suspended by a thong around the neck. These pendants have small holes drilled through the top and some simple carvings. A bone gouge or scraper was also found. The bits of charcoal we saw probably dropped from torches used by the Indians when making the burials.

There are several other noted bone caves in the vicinity of the Cumberland Bone Cave. In Bushey's Cavern near Cavetown, Maryland, remains of twenty-five different species of vertebrates were found, twenty of them similar to the ones from Cumberland. Bushey's is the oldest known cave in Maryland, having been mentioned in print as early as 1748. Originally, the cave was said to be so large that a thousand people could stand in the entrance. By 1925 quarriers had destroyed most of the cave, and the remainder has since collapsed; what is left today is too dangerous to explore.

Another bone cave that has since been quarried away but which yielded important fossils was one situated near Frankstown, Pennsylvania. Thirty-seven species, three of which were new to science, were recovered. Fifteen of the Frankstown species are similar to the ones from Cumberland. Most of the bones from Frankstown Quarry are now in the Carnegie Museum.

Along the Delaware River near Riegelsville, Pennsylvania, is the pitiful remnant of the once beautiful and important Durham Cave, site of both anthropological and paleontological excavations. Kitchen middens (the remains of food and broken utensils thrown into fireplaces) containing burnt and cracked bones had been found near the entrance. Undoubtedly the cave was a storeroom for a Lenape Indian village. The bones of over thirty different species of animals were unearthed and reported by various authors.

Perhaps the most famous of all bone caves in America was the Port Kennedy Bone Cave, near Norristown, Pennsylvania. It has been obliterated by quarrying operations, but between 1870 and 1900 over six hundred tons of bone material and clay were removed under the auspices of the Academy of Natural Sciences of Philadelphia.

Dr. Edward D. Cope, one of the great pioneers of paleontology, described fifty-three different species from these remains. Eighteen of the species were new to science, and most of the others represented creatures long extinct. Over half of the Port Kennedy animals are identical with the Cumberland remains, corroborating our picture of Ice Age and pre-Ice Age life.

Earliest Americans

Until the discoveries in Sandia Cave, New Mexico, Folsom Man, who lived 10,000 years ago, was our earliest citizen on record. Sandia Cave provided evidence of another, more ancient and mighty hunter who tackled mammoths and ground sloths. The cave was his home and the remains found concentrated therein tell more about him than a dozen scattered, non-cave camp sites.

One such site in Nevada, recently studied by an excavating party from the Southwest Museum, yielded charcoal 23,800 years old according to radiocarbon analysis, but that fragment of knowledge adds little to the picture of Early Man.

Dr. Hibben began his archaeological researches in New Mexico in 1934, and continued them while working for his doctorate at Harvard. Interrupted by his wartime service as a Lieutenant Commander in the United States Navy, his researches were resumed in 1945. Professor Hibben is on the staff of the University of New Mexico and during the summer directs the Archeological Field School.

FRANK C. HIBBEN

Natural caves were among the earliest shelters used by our remote ancestors. It was in a series of caves at Chow Kow Tien, near Pekin in north China, that the now-famous skulls and skeletons of Pekin Man were discovered. Pekin Man is one of the earliest and most important types of fossil humans known. In Africa, too, and in Europe, the bones or the tools of Early Man are found in caves and beneath the overhanging roofs of rock shelters.

Cave homes of early types of humans have been found in the New World also. But both North and South America have long been forced to take back seats when the question of the antiquity of man has arisen. Our predecessors in Europe, who killed the mastodon and grappled with the cave bear, are well known, and every grammar

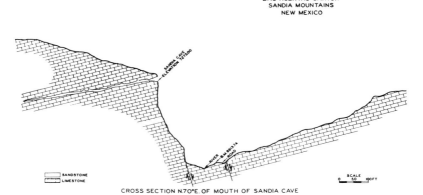

CROSS SECTION N.70°E. OF MOUTH OF SANDIA CAVE

school child is familiar with their appearance and their tools. The New World, however, has not been able to boast of 50,000 or 100,000-year-old inhabitants with hairy skins and protruding jaws. Instead, we on this side of the Atlantic have to be content with a moderate background of antiquity and with human beginnings derived second hand from the Old World by means of migrations across the Bering Strait.

It is for this reason that the discovery of a cave habitat of ancient hunters in New Mexico has been followed with much interest by anthropologists. The ancient American artifacts found here give promise of belonging to the earliest men yet found in North or South America, comparable in age with some of the European oldsters.

The discovery of this cave home was made by accident, as most scientific discoveries are. Kenneth Davis, a student from the University of New Mexico, was spending a weekend exploring caves in the vicinity of Albuquerque, New Mexico. The spot he had picked for his exploring was in the Sandia Mountains.

On the Monday morning after this weekend of exploring, Davis brought back to the museum at the university a cigar box full of treasures he had collected in one of these caves. These objects were nothing to get excited about in themselves—there was an arrowhead, a piece of a prehistoric coiled basket, part of a deer antler cut into a tool, and a few fragments of pottery. This kind of evidence can be found in practically every cave in the region. The pottery was Pueblo in type, probably not more than six hundred years old at the most.

This collection of oddments of moderate antiquity was, however, interesting enough to suggest further exploration of the cave. Accordingly, in the next week, several archaeologists visited the site. Actually, there were seven caves in a group on the east wall of Las Huertas Canyon in the north end of the Sandia Mountains. But only one of these caves was of any size, and it was more of a tunnel than a cave in the usual sense, leading back into the face of a cliff for over six hundred feet, but averaging only about twelve to thirteen feet in diameter. At several places throughout its length, pieces of rock had fallen from the ceiling and were piled almost to the roof. It was only by crawling, and in places by actually sliding on one's belly, that the back of the cave could be reached at all.

In the dust and debris of the cave floor one of the archaeologists scuffed up a bit of bone—a piece long and curved and sharp on one end. This was the claw of a giant sloth which had ambled over these limestone hills some 10,000 to 20,000 years ago. Discoveries in other sites of the Southwest had led to the belief that man existed at the same time as the giant ground sloth. Indeed, there was some evidence that ancient man had killed and eaten these animals when he first came to North America.

We know the great ground sloth well from other scattered discoveries. He was a huge, clumsy animal about the size of a large bear, and had long, yellow hair. The sloth's claws were so long that the animal walked on the back of his paws, with his claws turned underneath. The giant ground sloth was herbivorous and ate only leaves, which he raked from the branches of shrubs and trees with his long claws. Lumbering and slow as he must have been, the sloth represented some 1,000 pounds of meat for any hunter of the time. But the giant sloth has been extinct for 10,000 years.

Since this initial discovery, Sandia Cave has been the scene of much activity. Flickering torches and acetylene lanterns have sent shadows dancing in gloomy corridors where the sloth laired in bygone ages. The chambers and passages of the cave have reverberated with the clang of sledge and pick and the rattle of wheelbarrows.

Gradually there has come to light a human story which could be pieced together from evidence gathered during the four years of excavation in this cavern home. Here there was a hearth with the blackened fragments of a camel jaw beside it, though many thousands of years have elapsed since the camel, which originally evolved in the New World, was native to the Southwest. Dart points were scattered among the broken bones of animals which had been killed, perhaps, with these same points. Scrapers and knives of flint and bone splinters broken for the marrow gave evidence of Sandia hunters and their domestic life. A cave man of ancient America gradually took form.

This, in itself, is not remarkable, for signs of a very early type of American, who hunted now-extinct mammals and who lived during the rainy period just after the last glaciation, have been found before. Of these early Americans, one of the most famous is the so-called Folsom Man, named from the little town of Folsom, New Mexico. Folsom Man was a hunter who ranged up and down the foothills of the Rocky Mountains, pursuing a peculiar type of bison or buffalo now long extinct. This man left traces of his passing in a distinctive type of javelin point which looks like a short bayonet with a groove running up either side. These Folsom points have been found from Saskatchewan to Texas and for the last decade have been considered the earliest evidence of human habitation in the New World.

At first it was thought that the Sandia hunter was only another phase of the Folsom, a variation perhaps, or another tribe. Four years of digging have, however, produced conclusive evidence of a group of men who hunted the green hills of New Mexico long before even Folsom Man. These were contemporaries of the mammoth and the mastodon, and of the American horse and the camel and the savage carnivores who preyed upon them. Knowledge of these earlier animals is not founded on guesswork, or even on clues, but is based on the science of stratigraphy. Stratigraphic evidence relies on the long-known fact that he who is first gets in on the ground floor. Thus, if we dug in the city dump in New York, we should find the relics of the Gay Nineties buried below the more modern remains of our own era. We should correctly conclude that the battered remains of the horse car represented an earlier vehicle than the automobile which lay above it. The joker in stratigraphy is, of course, the difficulty of finding remains which overlie one another. If, instead, the deposits are side by side, no matter how primitive or advanced one or the other may be, it is difficult to establish conclusively the precedence of one over the other.

Sandia Cave, happily, is well stratified. On the top surface throughout the cave is a heavy deposit of dust, bat guano, and broken rocks fallen from the roof. Mingled with the dust at the front of the cave are fragments of pottery, baskets, and yucca sandals of the Pueblo era. Below the uppermost layer of dust is a crust of stalagmitic material some three to six inches thick. This records a period when the cave was wet and when water containing calcium carbonate in solution dripped from the ceiling and spread out over the cave floor. The calcium carbonate was deposited in a sheet as the water evaporated. Under ordinary conditions, perhaps only a fraction of an inch of this material is formed in a century. More important than the possibility of its indicating age in the cave is the effectiveness with which it sealed in the deposits below it, and so prevented any mixing of recent deposits with the ancient material which lies beneath. Thus the remains of extinct mammals and the fragmentary clues to the cultures contemporary with them were completely enclosed and lay unaffected by disturbing influences until the sledge and the pick of the archaeologist broke through.

The geological epoch just prior to our own is called the Pleistocene. It was characterized by the formation of great sheets of ice in continental glacier form in both North America and Europe. The fauna of the Pleistocene is extremely distinctive, insomuch as most of the species then extant are now extinct. Wet periods accompanying these glacial times made great changes in areas now dry. The mammoth grazed amidst plenty on slopes which are now barren and rocky. The stalagmitic crust in Sandia Cave is evidence of one of these wet periods, when the hill above the cave was deluged with rains and snows. All material below this crust is Pleistocene in date. No mammal bone or fragment of Pueblo culture occurs below this level.

Immediately beneath the calcium carbonate capping of the cave lies a thick layer of debris, dirt, and bone fragments, consolidated into a homogeneous mass by the same material as that of the capping above. This consolidated material is a great, flat hasty pudding mixed with ingredients which were lying around on the cave floor in the late Pleistocene. There are fragments of bone and teeth, representing the garbage piles of beasts and men. Pieces of rock of all sizes are covered with dust, blown or carried in from the cave entrance. Chips of flint, scrapers, and points are fixed in this material as though liquid cement had been poured over the whole to make sure nothing moved from the position it had assumed. Most

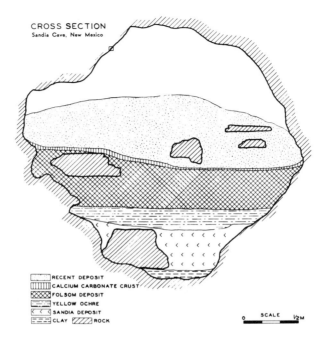

interesting is the fact that the projectile points in this material are of the true Folsom type. Evidently the Folsom bison hunters occasionally used the cave when on trips in the vicinity, and lost or left their implements there.

Beneath the Folsom layer lies a thick deposit of yellow ochre stratified in laminae and evidently water-laid. This, then, represents an earlier wet period in the cave's history. At this time the drip and trickle of the cave waters were not disturbed by any inhabitants or visitors. The sloth shunned the mud and slime of the cave floor, and so did man, for there is no indication in the ochre layer of any disturbance or any casually dropped implement or bone.

Below the ochre and between it and the solid limestone of the cave floor are the most important deposits of the cavern. These were accumulated during a dry period, the first in the varied history of this abode. In these lowest levels there again appeared traces of grisly meals and fires, bone fragments, and flint implements. These latter are javelin points entirely different from the later Folsom variety. They are notched from one side, forming a single shoulder, in appearance very similar to flint points of the Paleolithic or Old Stone Age of Europe. That the Sandia points may be comparable with some of the earliest implements of Europe is a distinct possibility.

That Sandia Man lived, hunted, and died in the time of the mammoth and the sloth we now know. Yet, in order to satisfy a perfectly natural as well as scientific curiosity, we should like a look at his bones—if only one of his dead lies in a corner of the cave, and if only the rats have spared him sufficiently so that we can see his outline.

European remains of the Paleolithic variety were left mainly by types of men who had certain primitive characteristics. They walked with a stoop-shouldered, bent-kneed gait. Their heads were fastened on their necks far forward like those of gorillas; their jowls protruded and their foreheads receded. Such men as these have been eagerly sought for in the New World, from the time they were known to European scientists. So far, there has been no success. First, Folsom Man, and now Sandia Man, seemed to give promise of being this American link.

When Professor Kirk Bryan of Harvard University carefully plotted the stratigraphic sequence of wet and dry layers in Sandia Cave, he correlated these alternating periods with the advances and retreats of the last glacial epoch. The Wisconsin Glaciation, as it is called in North America, was the last major manifestation of ice on this continent. But even this glacial mass did not simply advance at first and then melt back in its final stages. The Wisconsin glacial ice made many minor advances and corresponding retreats. Although the glacial ice did not invade what is now New Mexico, these oscillations produced wet or rainy periods there, interspersed by dry epochs. By calculations of this sort, Professor Bryan deduced that the Sandia men killed mammoth and mastodon and dragged their bones back to Sandia Cave some 25,000 years ago.

Recent deductions of a different sort have confirmed Professor Bryan's geological timetable very closely. Scientists had discovered, even before the creation of the atomic bomb, that all living things, both plant and animal, contain within their bodies substances which are slightly radioactive. The most important of these substances is a form of radioactive carbon called Carbon 14. All animals and plants, in their lifetimes, absorb Carbon 14 from the atmosphere around them. When they die, however, they cease to take in the

Carbon 14. Inasmuch as this form of carbon is radioactive, it continues to degenerate at a fixed rate. It is by this mathematically calculated degeneration of radioactive carbon that scientists have been able to determine just how old some archaeological deposits may be.

Recently Dr. H. R. Crane of the University of Michigan has dated three fragments of mammoth tusk from the Sandia level of Sandia Cave by the Carbon 14 method. Dr. Crane found that this ivory contained Carbon 14 in an amount which indicated that the animals which had carried these tusks had died at least 20,000 years ago.

The University of New Mexico is presently engaged in excavating other evidences of these earliest Americans. Just to the east of the Sandia Mountains lies the Estancia Valley. The Estancia Valley was, during the Ice Age, a large lake around whose grassy edges grazed animal herds which the Sandia men undoubtedly hunted. The lake is unhappily now dry, and the animal herds gone these many thousands of years, but on an old beach level of Lake Estancia erosion has exposed a layer of elephant bones mixed with the charcoal of ancient cooking fires. The typical side-shouldered javelin points have been found—sure indication that Sandia men camped and cooked there. It was perhaps from this camp site on the shores of ancient Lake Estancia that the Sandia hunters moved up to stay in Sandia Cave and then returned to walk around the beaches of the lake.

Somewhere on these ancient shorelines, or in other caves in the Sandia foothills, we may some day find a skeleton of one of these early American hunters.

Hidden Skeletons of the Mother Lode

A unique legacy of cave lore left by the forty-niners and their followers has furnished modern cavers with a host of tantalizing clues. Reports on caves ferreted from the files of newspapers, rare books, and unpublished journals have sent California cavers into uncharted desert and mountain areas.

Most of the intensive investigation was carried on by Edward A. Danehy while he was research chairman of the Stanford University Chapter of the National Speleological Society. At the time, Dr. Halliday was chairman of the Southern California Chapter. Here he tells how a series of amazing tales, confusing descriptions of caves, and scattered collections of skulls and skeletons were finally combined into a plausible story, if not into an incontrovertible one.

WILLIAM R. HALLIDAY

In the western foothills of the Sierra Nevada ranges of California lies the Mother Lode country, world-famous for the gold-hunting forty-niners. Today the story of its gold is largely history, and its interest lies in other fields. Speleology is one of them.

This country is full of caves. The Mother Lode area can almost be bounded by Juniper Cave in Plumas County on the north, and on the south by Bower Cave in Mariposa County, near Yosemite National Park. These caves of the Sierra foothills, although not the largest in the state, are geologically quite remarkable, for their marble bedrock has been turned up on edge so that it dips precipitously, and often is fully vertical. Many of the caves are of great beauty, and two—Mercer Caverns and Moaning Cave—are developed as tourist attractions.

It is neither their geology nor their beauty which arouses interest in several of these caves, however. Nor do they claim any part in the golden story of the forty-niners. No important gold deposit has ever been found in these limestone caverns, though the gold-hunters seem to have investigated nearly every one of them. Instead, it is their incredible content of human bones which attracts the explorer and scientist.

Cave hunting in the Mother Lode country is different from that in most other areas. The wooded ridges and rocky valleys are not nearly so well known today as they were a hundred years ago. Some old-timers who still recall the location of caves near their homes are valuable sources of information. Unfortunately, limestone outcrops in this country are small, obscure, and often unmapped, so that it is difficult to know where to begin local inquiry. As a result, a new type of correlative speleology has been developed in California, employing exhaustive bibliographic search as a basis for field study. Many of the Mother Lode caves have been rediscovered by investigating caves mentioned in old newspaper clippings or in half-forgotten books in some library's research department.

Some of the caves mentioned in these early accounts are still completely unknown, and their relationship to other caves is obscure. Correlation of the old and often vague references with caves known today requires supporting evidence. The mere interpretation of the old description of a cave is in itself insufficient. Fortunately this is often not the difficult task it might seem.

Nearly all of the major Mother Lode caves were well known to the pioneers. To these men, whose very dreams were of the products of the depths of the earth, caves were an object of the most natural curiosity. Their cavern finds were promptly investigated and frequently were recorded in the local newspapers.

Many of these fragmentary reports were examined by pioneer California scientists and historians. The findings of John B. Trask, first state geologist; Henry G. Hanks, state mineralogist; J. D. Whitney; the two historians John S. Hittell and J. M. Hutchings; and Isaac W. Baker, comprised the first true speleological studies in California. Their records have helped the present-day speleologist to unravel many perplexing puzzles of the Mother Lode.

Surprising discoveries were made by these early spelunkers and speleologists. In Calaveras County, explorers frequently found human remains, sometimes in great abundance. Much more than just an occasional skeleton was discovered beneath a cave's trap-like opening. Some caves contained astounding accumulations—whole charnel houses of rotting bones.

At the time of the discoveries, some Indians explained that the piles of bones were the skeletons of victims of a great battle, several generations earlier, but others could offer no explanation. Sixty years later, ethnologist C. Hart Merriam was told by members of the Miwuk tribe that the bones belonged to the victims of Chehalumche, a cannibalistic, troglodyte giant. The Miwuks declared that Chehalumche's caves were held in such horror that their people would never throw a body into one or voluntarily enter them. Neither legend, however, is sufficient to account for all the skeletons.

In Cataract Gulch Cave, a small, vertical cavern, a single skeleton was found. The striking fact about this particular skeleton was a gold tooth in its well-preserved jaw. Obviously this was no prehistoric relic. An old-timer promptly recalled the disappearance, many years earlier, of a miner with a gold tooth and well-filled ore pouches, and the mystery of one of the skeletons was solved.

In most caves where Indian skeletal remains were found, the entrances were vertical, funnel-like traps. In Cave of Skulls, O'Neill Cave, Moaning Cave, Mercer Caverns, and Mercer's Indian Burial Pit, the location and position of the bones indicated that the bodies could either have been thrown in or have fallen in. In these caves, wherever more than a very few bones were present, beads or other ornaments were invariably found. Cave of the Catacombs, however, is essentially horizontal, and the many victims whose bones were found there could not possibly have fallen in. Furthermore, no artifact has ever been found in Cave of the Catacombs. If it is assumed that all the large skeletal deposits in this small area came from the same source, then it would seem strange that the ornaments would be stripped from bodies placed in only one of the caves. Nevertheless it might have happened.

The unraveling of the story of Cave of the Catacombs is a triumph of speleology. The cave was visited by members of the National Speleological Society in 1950, before the bibliographic studies were far advanced. The cave was owned by a Mr. Jerry Miller and at that time was known as Miller's Cave. There was no reason then to associate it with the Cave of the Catacombs, known only from an early report of Henry G. Hanks, which stated: "In August, 1881, I visited a cave then rediscovered in Calaveras County which con-

It was at this spot that many of the skeletal remains in Cave of the Catacombs were found. At many times of the year there is no water at this point. *Edward A. Danehy*

tained a large number of human bones. This cave, near Cave City, was named 'The Cave of the Catacombs.' In exploring it, names of visitors were found dating back to 1850, at which time there must have been another opening now unknown . . ." since the entrance had had to be enlarged to permit entry of the discoverers.

In 1950, the University of California Archeological Survey also visited Miller's Cave, and filed a brief, but similar, report. It read: "Miller Cave, situated in a canyon between Adobe and Dry Gulches, was found to contain numerous human bones, some badly broken, others complete. The bones littered the floors of several galleries and some were observed in a pool of water at the bottom of the cave. No artifacts were recovered."

Further bibliographic research revealed an account in the Sacramento *Union* of the 1881 discovery of the Cave of the Catacombs. Finally a lengthy, detailed description of the cave and its contents was discovered in the San Francisco *Call*. Included in this key article was a rough sketch map of the cave, and a list of signatures found on the walls.

Miller's Cave had two entrances, both small and partially blocked. Otherwise, a comparison of its map with the sketch of Cave of the Catacombs revealed a startling similarity, and the description of the latter and its skeletons left little doubt that these were the same cave.

In hope of discovering final, conclusive proof of the identity of the cave, the names listed in the *Call* article were compared with those in the predicted area within Miller's Cave by members of the National Speleological Society's former Stanford Chapter. Immediately, all but three of the names in the 1850 period were found, and the others may still underlie the host of later names and dates. There was every indication that the low, mud-choked second entrance, which at first was so puzzling, is open only intermittently, and this must also have been the case in the last century. The name Miller's Cave, accordingly, was officially removed from the list of California caves, and Cave of the Catacombs substituted.

This still left the problem of finding the source of the unornamented bodies. Were they the stripped bodies of war victims or of Indians who died of natural causes long before the origin of the legend of Chehalumche or before the tribal memory of the Miwuks? Could they have been criminals, buried alive as an Indian form of capital punishment?

Another explanation appeared in the *Call* article in 1881:

"There is a legend told by an old Indian woman who at one time lived with her people in Calaveras that might not be out of place, although of little value ethnologically.

"The Indian woman has the reputation of being more than a hundred years old, but is perfectly sound in mind. Through an interpreter, the reporter got the following horrible history of the new cave:

"'For hundreds of years the cave has been used as a prison by the Indians. Long before the white man came to this country, I was given to the Fathers at the Mission San José. I was educated at the Mission school. . . . After I had been at the Mission about a year, my father and mother came to see me. They told me that my oldest sister had been put in the cave for leaving her husband. I had been taught right from wrong, and told Father Pedro all about the cave;

how, that when a woman survived, she and her children were placed in the cave to starve to death with the prisoners. The fathers sent up to see about it, and found several women and children in the cave, and among them my sister. Some had just died, others were dying, and some were eating the raw flesh off the dead bodies. I went with the priest to show them where the cave was, for none of the Indians would tell. I was very young at the time, but I never got over it. When the fathers moved the rocks away from the door, it was terrible. After a long time the poor creatures were brought out, but most of them died and the others became raving maniacs. The fathers buried the dead and sealed the cave up. After we returned they sent out among the Indians warning them against using the cave again, and I don't think they ever did.'"

All the findings in Cave of the Catacombs could be accounted for on the basis of this legend. Although the story might well have been pure invention of an imaginative reporter, as were the mythical Cyclopean Cave in Colorado and Magnetic Cave in nearby Amador County, equally strange legends have been proved true in a number of California caves. Samwel Cave, for example, is also called the Cave of the Lost Maiden. Digging there, a paleontological party from the University of California not only unearthed a rich deposit of extinct animal remains, but also found the mummified body of an Indian girl at the base of a deep pit, exactly as related in an Indian legend of the McCloud River area. The Catacombs story, however, was dismissed quite bluntly in 1884 by Dr. Hanks: "The sensational articles," he wrote, "which appeared in some of the papers at the time, to the effect that this cave had been at one time a prison, in which men, women and children had been driven to perish by starvation, were wholly without foundation."

Furthermore, the tale is in conflict with that of Chehalumche. It is quite probable that the actual facts may never be known, but it is more difficult for modern speleologists, in retrospect, to dismiss the story related by the alert *Call* reporter than it was for Dr. Hanks.

The history of Cave of Skulls, another Mother Lode cave, is similar to that of Cave of the Catacombs, at least up to a point. Unfortunately, there are peculiarities in its history, and in that of nearby Moaning Cave, which remain puzzlingly obscure even today. Cave of Skulls had not been known under a different, modern name, but its location had almost been forgotten.

HIDDEN SKELETONS OF THE MOTHER LODE

This scene along the Stanislaus River, in the Sierra Nevada foothills in California, is typical of the jagged, rough country in which the Mother Lode caves are found. Nearby is Cave of Skulls. Moaning Cave is about two miles to the west. *Edward A. Danehy*

The numerous detailed references to Cave of Skulls in archaeological and other studies of the last century led to its rediscovery by Edward Danehy and other members of the National Speleological Society. The vague descriptions of its location indicated that the cave should be found somewhere between Moaning Cave and the Stanislaus River, two miles to the east, but in this jagged, densely wooded country, this was of little help. Addison Carley, present owner of Moaning Cave, had visited a cave in Skunk Gulch many years ago which could possibly be Cave of Skulls. Through Carley, the ownership of the cave was traced and the fact that skulls had been removed from it prior to 1872 was ascertained, thus correlating the cave with the one known so many years earlier.

The rediscovered cave was entered through a natural shaft thirty feet deep, and was found to be principally a single, steep room about forty feet in slope length. Reached by a ledge on one wall was a narrow opening to a smaller, sinuous chamber. It is not at all a large cavern, but the large number of skulls and other relics originally strewn along the lower slopes of its chamber aroused widespread interest.

Nearby Moaning Cave is very different, although it, too, is a vertical, trap-like cavern. Its modern entrance leads to a series of zigzag, fissure-like passages, abruptly descending sixty-seven feet. An awesome spectacle then bursts upon the visitor, who finds himself at the top of an enormous room at a point from which its floor, a hundred feet below, cannot be seen. A few feet away and overhead is the opening of the nearly vertical original entrance, through which the first explorers and the scattered bones below entered the cave. Leading down from the base of this large chamber is a minor series of passages and small rooms, which reach a further depth of forty-two feet, and finally a pit leads to a small chamber at a total depth of 254 feet.

It would seem difficult for these two caves to be confused with each other. Prior to the construction of the massive spiral staircase which now leads to the base of the dramatic Big Room in Moaning Cave, descent was a formidable exploit. In contrast, Cave of Skulls can be entered without great difficulty. Nevertheless, the occurrence of human remains in two nearby pit caverns, and the uncertainties of nineteenth century frontier journalism, have combined to create confusion between the caves.

It is certain that Moaning Cave was known as early as 1851. The widely disseminated description of Solomon's Hole or Ossiferous Cavern by geologist John B. Trask was too exact to leave any doubt that Moaning Cave was being depicted. While Dr. Trask does not specifically state that he himself made the hazardous descent, the clarity of his account proves that he must actually have negotiated the one-hundred-foot drop to the floor of the main chamber.

Two years later an account of the discovery of an almost identical ossiferous cavern appeared in *Alta California,* a noted San Francisco newspaper of that period, under the title "Extraordinary if True." It related a descent of more than three hundred feet made by a party of French miners in a cave near Vallecito. A week later, a more conservative, but still third-hand, report was translated from a story in *L'Echo du Pacifique,* a French language newspaper, which reduced the size of the cave considerably. It described a thirty-foot vertical descent at the entrance, followed by a forty-foot slope, at the base of which was "another," nearly perpendicular drop, leading into a small opening and then to a large cavern room.

There is no counterpart of such a cave known in the Vallecito area today. Although some have interpreted the second vertical descent to be the one-hundred-foot free drop in Moaning Cave, study of the context of the article does not confirm this view. It is of interest to note that if an equally unremarkable ascent is substituted for this descent, and the size of the following chamber is shrunk quite drastically, a perfect description of Cave of Skulls is obtained. While there is no definite proof for any such assumption, the discrepancy could have occurred readily in at least three ways: poor description of size by the French narrator, Mr. Alhinc; difficulties in translating the original reference, now lost; and journalistic license, which must frequently be reckoned with.

When the problem seemed most puzzling, an invaluable discovery was made. This was an unpublished 1853 account by Isaac W. Baker, containing a description undeniably that of Cave of Skulls. It refers to "a remarkable cave [which] was accidentally discovered by a party of Frenchmen engaged in mining in this [Vallecito] vicinity during the summer of 1853. An exaggerated account, or rather report of its discovery was published in many of the California papers." Included was a fine sketch of the cave which left no doubt as to its identity. Finally, another *Alta California* reference, dated one day later than that translated from *L'Echo du Pacifique,* locates the correspondent's route along the old road almost alongside Cave of Skulls, stating: "At the top of this hill (ascending from Abbey's Ferry) is the newly discovered cave in which a number of human skulls have been found."

Moaning Cave is by far the largest and most important of the major Calaveras sepulchers. When Dr. Trask explored it, he probably descended through the more vertical old entrance passage. Accurately, he judged the descent from the surface to "the jumping off place" as seventy feet, and the free descent into the great chamber as nearly one hundred feet. Consequently, his error in overestimating the maximum width of the chamber as three hundred feet, rather than about half that, will be pardoned by all who have been there. The feeble torches of Dr. Trask's party could not possibly have lighted the opposite walls of the immense chamber adequately enough for accurate judgment of size.

Dr. Trask wrote: "A large mound occupies the center of this room, fifty feet in height and seventy feet in diameter, composed of loose stones and earth, that has washed in from the top, and contains gold." (Today, the cone has been flattened to make a platform for tourist observation.)

Describing a sketch which no longer exists, he stated: "Behind the figure, sitting by the fire (at the bottom of the room) you will notice a triangular space, in the distance forty-six feet on the scale. This is the aperture to the next chamber, below and directly under the first; it cannot be shown on the plate. The vertical depth of this room is one hundred feet, and is composed of fragments of the rock, forming the cavern; in this chamber the most interesting feature of the whole presents itself, which was the appearance of portions of a human skeleton. On a large flat rock, on one side of this room, lay a portion of the skull. . . ."

Later he mentions explorations that had been made to a depth of 450 feet, beyond which two unplumbed pits still were said to remain unconquered.

Some speleologists believe that the latter part of this quoted description was meant to be that of a second enormous chamber, still deeper than any now known, or perhaps an exaggerated, second-hand account of the small lower passages, where a couple of skeletons have been found, and which Dr. Trask probably did not visit. On the other hand, this entire section of his narrative seems so clearly to pertain to the enormous main chamber that any other interpretation is hardly justified. Certainly, persistent exploration and excavation have failed to bring to light a lower room of any great size.

Excavations at the base of the great chamber in search of a lower room nevertheless proved rewarding in an unexpected manner. Dr. Trask's conjectures concerning additional remains and the great antiquity of the skeletal material have been more than vindicated. During the excavations, great numbers of human skeletons and their fragments were exposed. Some were intact, more broken. With the bones were found remains of torches, bows and arrows, beads made of olive shells, abalone shell pendants, and other ornaments. In 1950, the University of California Archeological Survey uncovered and studied the remains of eleven skeletons, two of which were those of children. The artifacts were indeed not the work of recent Indians, but were of the type in use from 1500 B.C. to A.D. 500. Many of the bones and shell ornaments lay beneath a thick flowstone coating, buried deep in red clay.

When the Santa Barbara Museum of Natural History took over the project, skeletons of another "6 to 10 individuals" were found while another part of the room was being excavated. A half-dozen skeletons had been removed from Mr. Carley's original excavations. From the 25 per cent of the cave floor so far excavated, it is possible to calculate that the remains of about one hundred individuals rested in Moaning Cave.

The great age of these bones is attested to by the innumerable layers of the stalagmitic flowstone which surround some of them. One human femur was imbedded beneath three and one-half inches of flowstone, in which over 1,400 distinct rings were noted.

Unfortunately, travertine layers are not as accurate guides to age as are the annual rings of trees. In some caves, it might be safe to state that speleothems build up two layers per year. On the other hand, most speleologists believe that local variations in precipitation so change the rate of deposition in any given cave, or even in a single stalactite, that calculations based upon flowstone thickness are not usually reliable. In discussing the relative age of human bones and those of an extinct ground sloth found close together in nearby Mercer Caverns, anthropologist William J. Wallace of the University of California Archeological Survey was unwilling, in 1951, to assume that the skeleton of the ground sloth was older than the human bones merely because a greater thickness of flowstone coated the bones of the ancient beast.

There is no doubt that these human remains in Moaning Cave are of considerable antiquity. Radiocarbon dating has not, however, been applied to the skeletons. Until this is done, the correlation of the ornaments found in the cave with those in use elsewhere about 2,500 years ago seems an acceptable dating technique.

Approaching the dating problem in another way, however, Phil C. Orr, curator of the Santa Barbara Museum, has studied microscopically the mineral deposits on nails that have remained in the cave for twenty-nine years, since the building of the spiral staircase. He found that flowstone thicknesses from 0.24 to 1.0 mm. had been deposited on the nails. On this basis, he calculated that an average period ranging from twenty-nine to one hundred twenty years was required for the deposition of one millimeter of flowstone in Moaning Cave during modern times. Quite similar figures were obtained from a stalagmite-encrusted miner's pick which was known to have been in the cave for seventy years.

A maximum of 420 millimeters of flowstone covered the remains of the victims, who, Orr believes, apparently fell or were thrown into the cave at widely separated intervals, rather than en masse. *If,* and this is the major assumption, the average rate of deposition over the last several thousand years was the same as that over the last twenty-nine (or seventy) years, and *if* all possible flowstone was deposited on the nails, then the deposits overlying some of the bones required at least 12,180 years to form, and perhaps as much as 50,400 years!

Orr's minimum estimate of 12,000 years is approximately equal to the age of the most ancient remains of man in America verified by the radiocarbon calendar. His maximum age estimate of 50,000 years is a stirring challenge to the traditional concepts of American anthropology. As he points out: "There will be many eyebrows raised at both figures, but let us point out that from Carbon-14 determinations we know that man had already reached various parts of America by 12,000 years ago, that man in Nevada was trading with the Pacific Coast for sea shells as long as 9,000 years ago, and that hundreds of different cultures had developed through America to the southern tip of South America before the time of Columbus." The stratigraphic researches of Professor Kirk Bryan indicate that it is possible that a prehistoric hunter known as Sandia Man lived in New Mexico as long as 25,000 years ago.

Perhaps the most serious objection raised against Orr's theory is the fact that in most California caves, present calcite depositions appear to be occurring at a small fraction of past rates. In many such caverns, huge, ancient speleothems stand dry, dusty, and discolored by ages of quiescence, for cave formations no longer grow when water ceases to drip downward through the overlying rock. In some cases, the water which now drips is actually redissolving part of the speleothems. This is not the situation in Moaning Cave, where some active speleothem development is still in progress. It indicates, however, how difficult it is to find conclusive proof in determinations of this nature. Furthermore, if the smooth, lightly oiled surface of the nails repelled deposition of *any* flowstone, Orr's figures must be revised downward.

Proof is still lacking for this fascinating theory. Regardless of its outcome, the debate will have been worthwhile. If Orr's estimate is correct, all the theories on the antiquity of man in North America may have to be radically altered. If not, from the intricate research necessary to prove or disprove his concept will come new tools of speleology valuable in deciphering other, equally important mysteries. Such a challenge is fully in the bold tradition of the Mother Lode country, which once yielded treasures of another kind, and where even the most unsuccessful gold-hunter contributed to the building of California and the nation.

Bats and Bombs

Man has always been attracted by the spectacle presented by huge flocks of migrating birds or great nesting colonies. The saga of the caribou migration and the suicidal march of the European lemmings have intrigued scientist and layman alike.

No less astonishing are the evening flights of bats from caves in the American Southwest. Only recently have efforts been made to calculate the size of the bat colonies whose living quarters are the walls and ceilings of spacious underground chambers. The investigation of bat caves which Mr. Mohr describes had a military purpose and was incidentally the most extensive survey of caves ever carried out.

Charles E. Mohr

A hundred years ago the fabulous colonies of Passenger Pigeons and the hordes of Mexican Free-tailed Bats were America's greatest concentrations of wildlife. Today one is gone forever, and the other is greatly reduced in numbers.

So tremendous were the migratory flocks of "Wild Pigeons" that the passing birds darkened the sky for hours, and even for days. Along the Green River, near Mammoth Cave, Kentucky, in 1813, John James Audubon described a flight which lasted three days. No one knows how many birds took part in such a migration. Audubon calculated that in just three hours, more than one billion birds passed overhead.

During the nineteenth century the big doves were subjected to the most savage wholesale slaughter. Dozens of freight carloads a day were shipped from Wisconsin, Michigan, and other places where the birds concentrated. This destruction, together with other more obscure factors, led to a rapid decline in numbers. By the end of the

The Free-tailed or Guano Bat, *Tadarida*, is the most abundant species in southwestern United States. This was the bat used for the bomb project. *Charles E. Mohr*

century, they were almost extinct. No one has seen a live Passenger Pigeon since 1914.

The Free-tailed Bat, *Tadarida mexicana*, is still abundant in certain parts of the Southwest, yet its numbers have been much reduced. It has vanished from many caves where great colonies flourished in the past.

Like all bats, *Tadarida* has suffered from man's ignorance and intolerance. Lack of knowledge about these shadowy denizens of the night has led people to believe all sorts of fearful things about them. Superstitions abound—that bats become entangled in women's hair, that they are swarming with loathsome parasites. And mistaken ideas—that they are blind, that they are a kind of strange bird.

None of these notions is true. Bats are too skillful in dodging to entangle themselves accidentally in a lady's tresses and never do it deliberately; they are as clean as other mammals or birds. They have well developed eyes, though they depend on them less than on their hearing ability; and being mammals, they give birth to living young. Bats devour incredible numbers of insects, and the vast quantities of bat droppings called guano which have accumulated in large bat caves have been extremely valuable sources of nitrates for gunpowder and fertilizer. Yet tens of thousands of bats have been shot, and many have been clubbed and burned to death.

Nevertheless, man has played a smaller part in reducing the population of these winged mammals than he did in the extinction of the Passenger Pigeon. It may even be argued that the bats' decline antedated man's interest in the caves in which they roosted, and that natural forces have taken the leading role.

In New Cave in Slaughter Canyon, within Carlsbad Caverns National Park, New Mexico, I have seen bushels of bat bones under layers of travertine inches thick. It is not known whether these fossil skulls and scattered skeletons were the result of a single catastrophe or whether they accumulated through centuries of time.

It is possible that the bats hung so close to the cave entrance that they were "caught napping" by an unusually early "norther," a cold air mass such as sometimes sends the temperature plunging to 30 degrees below zero, even in New Mexico. While most *Tadarida* probably fly farther south for the winter, it is quite conceivable that once in a hundred centuries such a lethal cold wave may have struck before they could migrate. But no matter how or when they died, the bats certainly are gone from New Cave now.

They have vanished, also, from scores of other caves where huge guano deposits attest to their former prevalence. Without doubt these insect-eating bats flourished in an earlier period when the climate in the Southwest was more humid. But with the coming of more arid conditions, populations of flying insects probably were greatly reduced. As food supplies diminished, whole colonies of bats must either have perished or moved southward to more tropical areas abounding in insects.

It is believed also that bat colonies numbering millions were destroyed by fire. One early cave explorer who visited a large number of caves declared that there was no cave in the Southwest that had not been afire in some remote time, since entire cave floors were often covered with compacted ashes, sometimes fifteen feet thick. These fires resulted from spontaneous chemical combustion, he wrote, generated by the heat of the decomposing guano, which, in time, becomes converted largely into nitrates and nitrites, or saltpeter.

Yet some fires can be traced to man's activities. One cave along

Comanche Creek in Texas is known as Blow-out Cave, because of an accident which occurred in 1856. A hunter who tracked a bear to the cave built a fire in the entrance, hoping to drive out the creature. A tremendous explosion followed. No sign of the bear, or of the hunter, was ever found, but without doubt they fared as badly as any bats which may have been living there, for the saltpeter deposit is said to have burned for two years.

Fire in a cave produces what ranchmen call a "smoke hole," which when seen at night must be truly an awesome spectacle. One writer described a burning cave, situated on a mountainside, whose "stillness and solitude, with the tips of a few trees silhouetted, gave it the appearance of a cyclopean eye." Undoubtedly many fires were set deliberately, but in any case, the suitability of a cave for bat roosts was affected, whether the bats were present at the time of the fire or not.

The explosion at Blow-out Cave taught many people that saltpeter would burn, a well-known fact in the East after the War of 1812, when many southern caves were mined for earth from which the nitrates were extracted for making gunpowder.

During the Civil War a powder factory was erected a few miles below San Antonio, and bat guano and niter-bearing cave earth were carted there. Other factories were located close to major bat caves, such as Beaver Creek Cave, about one hundred miles north of San Antonio.

A survey of central Texas caves after the Civil War indicated enormous supplies of guano, with at least a dozen caves capable of producing enough niter-bearing earth to justify the building of a factory nearby. Many more contained huge guano deposits but were inaccessible or too difficult to mine. Most of these caves were within sixty miles of San Antonio, yet the bat caves of the Edwards Plateau, in central Texas, form a much more extensive cave area—250 miles long.

One of the biggest saltpeter mining operations was at Verdi, or Frio Cave, as it is now known. Located in Uvalde County, it is at the western end of this cave region. This was a tremendous operation. A wagon road made a half-circle around the 600-by-300-foot entrance chamber, ran down along a 20-foot-high trestle into a second room of 1,000 by 600 feet, and then continued along the wall of an

Week-old, naked bats clinging to ceiling of Ney Cave. They will be able to fly at the age of three to four weeks. Adult bats can be seen at upper and lower left and center. *Texas Game and Fish Commission*

equally large third chamber. At least a thousand tons of saltpeter were leached here, and operations continued into the twentieth century.

Less well known during the Civil War, but destined to become famous nearly a century later, was Ney Cave, halfway between Frio Cave and San Antonio. It was named for the Ney brothers, who owned the ranch on which the cave was located; the family had carried on saltpeter operations for generations. A flat, oval opening in the side of a low hill permits a climb down a 45-degree slope for about two hundred feet into a medium-sized chamber, the walls and ceiling of which are completely blanketed with bats in summer. Every square foot of the cave seems covered with bats. An upper level can be reached from the entrance chamber with some difficulty. To facilitate the removal of guano from this "top floor," a thirty-six-foot shaft was bored to it from the surface.

In Ney Cave, as in many of the bat caves, it is possible to walk for a thousand feet without stepping off guano, but one's footing is always uncertain, because beneath the guano are numerous large, irregular limestone blocks. During summer, when bat droppings are soft and fresh, an explorer often sinks knee-deep.

Fifty years ago a writer vividly described his impressions: "In texture the guano varies from a softish mass, into which one sinks readily and disgustingly, to a brown dry material that can hardly be compressed into a ball in the hand. Now and then it appears somewhat moldy, and this may be due to its being very old." His departure was rather abrupt after he had become badly mired in the guano, for, he said: "As I was quite alone, and the light decided to go out just then, I followed its example."

The strong ammonia fumes from the excrement are enough to deter most explorers and, at times, may even cause the bats to seek other quarters. Insects far outnumber the multitudinous bats. In Carlsbad Caverns, where I climbed over mountains of fresh guano in the bat chamber, far from the tourist route, I found myriads of beetles, insects, and other tiny, joint-legged creatures. Dermestid beetles by the millions were at work consuming the fur and flesh of dead bats and leaving exposed the complete skeletons of their prey. Bat parasites swarmed over everything, especially the rock surfaces recently vacated by bats. I saw strangely flattened flies and quarter-inch-long bugs closely related to bedbugs. They secrete themselves in cracks in the limestone rock. Tiny red mites were everywhere.

The constant rain of guano and parasites is enough to discourage even the most ardent of bio-speleologists. When one is "crawling" with parasites, the knowledge that they will not *live* on humans is small consolation. Certainly few spots that one can imagine are less inviting than a major bat cave in summer. The saltpeter miners knew this and worked almost exclusively in the late fall and winter months.

Today, as in the past, most ranchers who own bat caves are interested in protecting the guano-producing bats. Fertilizer rather than gunpowder is the end-product of present-day guano mining, and prospectors still wander through the Southwest in search of caves holding deposits of sufficient size to make mining operations profitable.

In Texas the Free-tailed Bat is a protected animal. The Texas Legislature in 1917 passed a law which makes it a misdemeanor (with a fine of not less than $5 or more than $15) "to wilfully kill or in any manner injure any winged quadruped known as the common bat."

The Texas law was the result of a prolonged educational campaign by a San Antonio physician, Dr. Charles A. R. Campbell. That a large bat colony was an economic asset, the Texas doctor demonstrated by collecting and selling the guano for fertilizer. During World War I he organized the "Texas Bat-Cave Owners' Association" which marketed substantial amounts of the valuable fertilizer.

Campbell carried out a number of experiments, one of which was designed to show the skill and speed of bats in finding their way back to their roost. He "tagged" 2,000 *Tadarida* with conspicuous white markings and carried them thirty miles from their cave roost. Immediately after turning them loose, he raced back to the cave in a "high-powered" car, covering the distance "in just fifty minutes."

"Within eight minutes," he wrote, "the vanguard appeared, then the large numbers, and began dropping from a great altitude and darting into the side entrance of the cave."

In investigating bat caves when the bats are active, there is the possibility of encountering what Pat White, the explorer of the Devil's Sinkhole, not far from Rocksprings, Texas, described as a "bat blitz." Accompanied by two other spelunkers and carrying a gasoline lantern, Pat noticed that the great colony of bats in Frio Cave was becoming quite active. Whole patches of bats would drop from the ceiling and peel off into the darkness.

"Each cluster as it came away flew clockwise around the room, pyramiding whisper upon whisper until a muffled roar surged through the enormous cavern as it slowly filled, as though by a great expanding whirlpool, with spinning bats.

"The bats on the fringe of the rotating circle began to fly into us, cling to us where they struck, and crawl upon us, demanding to be plucked away.

"Then they came too rapidly to be plucked or even brushed away. Many of them abandoned the spinning circle and dove straight at the gasoline lantern which was wedged in a niche from which they could not knock it. They struck like hail in a high wind."

By the time Pat realized that the lantern seemed both the cause and object of the attack, the three spelunkers were virtually buried beneath a mass of crawling bats. As he attempted to extinguish the lantern, he had to rake bats constantly from his face in order to see what he was doing. At last he got it turned off.

"In the welcome darkness we huddled together fighting the bats and fighting panic. But fortunately the battle was not extended. Almost immediately with the coming of the dark the bats receded toward the ceiling. And as we got rid of those which were already upon us they were not replaced. We waited till their sounds were high above us before venturing to shoot a flashlight beam into the darkness.

"Instantly the bats began diving at the light again exactly as strafing airplanes dive upon a target. They made it necessary to retreat from the chamber in stages—putting on the lights and moving and then waiting quietly in the dark before moving again.

"Later we found it possible to move without great trouble if only a single light be used and its beam not be cast upward."

Pat White concluded that the actions of the bats should not be construed as an attack upon the spelunkers. He compared the situation to the behavior of moths which are strongly attracted to a bright light. And like moths, many of the bats must have died as they hurled themselves upon the hot lantern.

No other spelunker has ever reported a comparable blitz, but, after all, few have ever attempted to explore a cave inhabited by millions of bats. Even a few thousand bats can create a "traffic jam" when they are suddenly aroused in a small chamber or passage. At times of such confusion their sound-navigation system is obviously ineffectual.

I have a vivid recollection of a swarm of bats that tried to get past me as I was stringing up a net in a passageway in Indian Cave, on the Holston River in East Tennessee. Many bats plummeted to my feet as they collided with each other and with me. Before I could remove the net, several hundred bats had become hopelessly entangled in it. I worked for hours until I had extricated all of them alive.

The evening emergence of hordes of cave bats is one of the most spectacular sights that anyone can witness. Carlsbad Caverns and many other caves owe their discovery to the bat flights that attracted attention and led persons to their entrances. In flat country, although even such large openings as the Devil's Sinkhole might pass undetected from a hundred yards away, the bat flights sometimes can be seen from a mile or more, creating a funnel-shaped cloud like an oncoming tornado.

Such flights are not overlooked by predatory creatures. Apparently a few bats "crash land" after aerial collisions in the entrance passageways, enough, certainly, to attract raccoons and skunks. Both Hognosed and Striped Skunks were observed by Jack C. Couffer while he was stationed at Ney Cave and at Bracken Bat Cave in 1943. He also saw similar carnivorous animals when he made a number of trips to Frio Cave and Blow-out Cave. Couffer found that the predators were unconcerned by his flashlight and continued foraging for bats. Evidently they could hear their prey hit the ground, since the fallen bats were pounced upon before they could recover and take off. A bobcat was seen in one cave.

Some predators consistently avoided being seen. Every morning the skinned and decapitated carcasses of twenty to thirty bats were found about the entrance by Couffer. The mystery of the predator which ate only skins and heads was never solved.

There is nothing mysterious, though, about the birds of prey that are attracted by the daily bat flights. Hawks and owls of several species gather outside the cave by mid-afternoon. The hawks circle and dive while waiting for the bats to appear. Watching them outside Ney Cave in August, 1938, naturalist Kenneth E. Stager noted six Peregrine Falcons or Duck Hawks "warming up" before the cavern mouth and screaming "as if they were calling the bats to come out."

Suddenly a compact mass of bats burst from the cave entrance, rose steeply, and formed a column fully fifteen feet in diameter, which leveled off at a height of about three hundred feet and disappeared to the east.

"The instant that the bats made their appearance, the band of falcons set to work," Stager relates. "Darting from above or from the flank of the column, the birds would cut into the onrushing mass of bats with talons set, and they seldom emerged from the opposite side without their prey held fast."

The big, streamlined Peregrines are so spectacular in their stooping dives and other forms of attack that ornithologists have journeyed to Ney Cave just to observe their tactics. A dozen trips were made by Alexander Sprunt, Jr., who describes several of the methods of attack:

"Not always do they follow typical falcon tactics. The stoop is, of course, indulged, but so was parallel flight with sudden swerves into and among the bats. Now and then a straight course at right angles to the stream was taken. Again, a bird would come under the bats and shoot upward, reach out with the feet and seize the prey. On one occasion, a Duck Hawk was seen to do this, and miss its strike, whereupon, in almost the same instant, it reached sideways with one foot and caught a bat. It just happened that the bird was in the field of a 9 by 35 binocular at the moment.

"On several occasions, after a bat had been taken, the bird made a meal of it on the wing. It fed exactly like a Swallow-tailed Kite, bringing the foot with the prey forward, bending the head back and down to meet it, and tearing the bat to pieces, discarding the wings which fluttered down like dark leaves."

The Sparrow Hawk or Kestrel, a smaller falcon, has been seen feeding on the bats, and so has the little Sharp-shinned Hawk. Generally, however, it is the large Red-tailed Hawk or the swift Cooper's Hawk which feeds most frequently on bats. I watched one Cooper's Hawk preying on Free-tailed Bats as they emerged from their cave roost near San Antonio one evening.

The hawk flew a set course, coming from just above the portal and flying in the same direction as the bats, which were appearing at brief intervals and in small numbers. I watched as the hawk made several successful sorties. In each case it came up swiftly behind the unsuspecting bat and grasped it before the victim could begin its familiar, evasive dodging. If the hawk encountered no bats, it swung sharply around at a distance of about fifty yards, flew back, and repeated the flight over the identical route.

I once saw a Barn Owl perched motionless on a ledge, obviously watching the flight, but perhaps waiting for darkness to fall before joining the attack. The only successful owl onslaughts reported were observed at Carlsbad Caverns, where Couffer watched a Great Horned Owl dive repeatedly at the flying stream of bats. Only about one out of every eight or ten attempts resulted in the capture of a bat.

Red-tailed Hawks are known for their soaring ability rather than for swift dives, yet they, too, are attracted by great flights of bats. Sometimes, says Sprunt, they "stoop" into the living river of bats "almost like a falcon, descending from above with half-closed wings and plunging through the bat stream to zoom upward with a bat in the talons. Again they will parallel the stream, then veer sharply into it and either emerge on the other side, or re-appear on the same flank, with prey. They seldom missed."

Few persons have witnessed the return flight in early morning. According to Kenneth Stager, it is made in the same manner as the evening flight in that "the incoming bats flew to a point several hundred feet directly above the cavern's opening and then volplaned downward and into the yawning mouth of the cave at a terrific rate of speed. By training one's eye to the top of the descending column, bats could be discerned approaching the diving point from all points of the compass.

"The mouth of Ney Cave opens as an oval doorway, ten feet high and twenty feet wide. This made it necessary for the descending bats to execute a sharp turn as they entered the cave. I found myself within feet of the rapidly descending column of bats. They were so close that the ripple of air on their wing membranes was clearly audible and much like the sound produced by air being rippled over rubber sheeting.

"Shortly after daylight, the Duck Hawks put in their appearance and immediately set to work securing their morning meal of bats. The capture of incomers was not as easy a feat as that of the preceding evening, largely because the incoming column was not as compact as the outgoing stream and was descending at a much greater speed."

Dramatic as these daily attacks certainly are, the hawk predation at Ney Cave is inconsequential. For the main flight here lasts for hours, from approximately five or six o'clock to midnight or even later, according to Sprunt. He estimates the total daily kill as not much over fifty, and the bat population at Ney Cave is the largest in the United States. Dr. L. S. Adams judges it to be between 20,000,000 and 30,000,000.

Dr. Adams should know because he led an exploring team which visited 1,000 caves and 3,000 mines in search of great concentrations of bats. Ney Cave ranked first in bat population, with Bracken Cave a close second, well ahead of Carlsbad Caverns, which had long been considered the country's greatest bat cave, since nearly 9,000,000 live there. Frio Cave and the Devil's Sinkhole complete the "first five" on his list.

The survey of bat caves was a preliminary step in a wartime project which came to be known as "Operation X-Ray." I talked with Dr. Adams a short time after the top-secret operation was first disclosed to the public. I knew him to be a fellow Pennsylvanian, a surgeon, with a variety of substantial inventions to his credit. He told me how he had arrived at the idea of bat-borne incendiary bombs—described in *Life* magazine as "one of the most extraordinary military operations ever conceived."

"On Sunday afternoon, December 7, 1941, I was driving near Washington when the news of the Japanese attack on Pearl Harbor

Dorothy Reville, of the New York Zoological Society, and Howard N. Sloane examining Cluster Bats, *Myotis sodalis,* hanging on calcite ridges along ceiling cracks in one of the wild caves in Mammoth Cave National Park. *Charles E. Mohr*

came over my car radio. Our navy was crippled. How could we fight back, I wondered. What offensive weapons did we have?

"I had just been to Carlsbad Caverns, and had been tremendously impressed by the bat flight. Now the thought flashed through my mind—couldn't those millions of bats be fitted with incendiary bombs and dropped from planes? What could be more devastating than such a fire-bomb attack? If dropped over Japanese industrial centers, fleet concentrations, ammunition dumps, or underground or other storage depots, the bats would seek shelter in inaccessible cracks and crevices above and below the surface of the ground and set off without warning a multitude of explosions and fires."

Wanting facts to bolster his idea, Adams set out for Carlsbad the next day to capture some bats and test them. Within a few weeks he was back in Washington with his bats and some startling facts.

Having ransacked the Library of Congress and museum libraries for information on bats and bat caves, he prepared an outline of his proposed bat-warfare project and sent it to the White House.

The idea appealed to President Roosevelt and his military advisors, and the project was authorized immediately, with Adams in complete command. As the first step, a small group of field naturalists, chiefly from the Hancock Foundation, University of Southern California, was chosen to make the search for the country's greatest bat concentrations.

"We visited a thousand caves and three thousand mines. Speed was so imperative that we generally drove all day and night when we weren't exploring caves. We slept in the cars, taking turns at driving. One car in our search team covered 350,000 miles."

The rigors of almost continuous high-speed travel were nothing compared to the exertion and hazards of the underground surveys. Untrained in safety techniques now in general use for all dangerous descents, the project members, including the middle-aged project leader himself, took their lives in their hands almost daily.

In the Devil's Sinkhole they descended two hundred feet on rickety ladders. Their deepest exploration was in an old mine at Quartzite, Arizona, where they explored a section nine hundred feet below the surface, using a haphazard collection of ropes. "We found bats in the lowest part of the mine and watched them fly straight up the shaft, nine hundred feet to the surface," Adams reported.

The deepest cave, and his choice as the most interesting, is the Devil's Sinkhole. Next to that in depth is Rose Cave, which apparently had never been explored before. When their ropes proved too short, the investigators gathered up hundreds of feet of barbed wire fence, and, with short sections of brush for rungs, improvised a ladder and reached the bottom.

While the bat caves were being surveyed, scientists from the University of California at Los Angeles, Southern California, Harvard, and the Massachusetts Institute of Technology were working on other phases of the project. They learned that a bat could actually fly with a bomb three times its own weight. The biggest bat found in the United States or northern Mexico is the Mastiff Bat, *Eumops*. With its wingspread of twenty inches, this bat was able to carry a one-pound stick of dynamite. The survey proved that Mastiff Bats existed in much greater numbers than naturalists had suspected, but not in sufficient abundance for the purposes of the project.

The more common Mule-eared or Pallid Bat, *Antrozous,* could fly with a three-ounce load but was not hardy enough for the rigors of bomb-carrying. After a dozen species were tried, the Free-tailed Bat, *Tadarida,* was selected. Although it averaged only one-third of an ounce in weight, this bat could carry a one-ounce bomb load. A trial run with dummy bombs was made in the War Department in Washington. As a small group of high-ranking military personnel watched in amazement, the bats navigated easily with their unnatural burdens.

Later the bats were returned to their cages, to be transported to the project headquarters in New Mexico. On the trip they were exposed to temperatures as low as 10 degrees below zero.

"They were frozen solid," Adams declared. "Yet when I reached New Mexico, the bats were alive. I opened the cages and the bats flew out, apparently no worse for their experience. That convinced me that this species was hardy enough to ship all over the world in cold storage. So we built an artificial cave, actually a huge refrigera-

Ney Cave entrance. Boxes stacked in front of the entrance were used to transport bats to a refrigerated storage room. *Jack C. Couffer*

tor. We kept hundreds of thousands of the bats in it, at a 40-degree temperature. This was essential because the bats would have speedily oxidized all their food reserves if allowed to remain long at normal temperature, and then they would have had to be fed or they would have starved to death."

After considerable experimentation, special bomb-like containers were designed. They held trays or fillers resembling the kind used in egg crates. Each container, carrying 1,000 to 5,000 bats, was

Bat with bomb attached. The one-ounce incendiary weighed three times as much as the bat, would burn for eight minutes. *Naval Photographic Service*

brought into the refrigerator room for packing. There small bombs were attached to the dormant bats, and each was placed in its own little cell, where a safety device provided against accidental detonation.

The one-ounce time bomb was slightly larger than the body of the bat and was attached to the loose skin over its chest by a surgical clip and a short length of string.

It was found that during a long parachute drop, enough time elapsed for the bats to become warmed and awakened. A simple mechanism opened the container at an altitude of 1,000 feet to allow the bats to fly away to secluded places of their own choice. Upon reaching some hiding place, the bat generally chewed off the string, then continued its exploration. When the small bomb finally exploded, it produced a twenty-two-inch flame which burned with intense heat for eight minutes, quite sufficient to ignite almost anything.

"Fortunately for the development of the project," said its leader, "Free-tailed Bats proved to be incredibly abundant. Our year-long survey of bat caves established the fact that Ney Cave was unquestionably our greatest bat cave. We calculate that we saw 100,000,000 bats in the caves and mines of Southwestern United States."

The United States Navy leased the four Texas caves with the largest colonies and assigned Marines from Corpus Christi to guard them. In October, 1943, screened enclosures were erected at the entrances of Ney and Bracken, so that the bats could be trapped as they attempted to emerge at dusk. The enclosures were prefabricated at Hondo Army Air Field and trucked to the cave entrances. Three big sections of screens at Ney and seven at the larger Bracken Cave enclosure could be opened, affording the bats a fairly clear exit or entrance. With such an arrangement the bat collectors could close the screens and capture 1,000,000 bats or more in one night if necessary.

Tests with live bombs demonstrated the effectiveness of the aerial "fire-bugs." A dummy village built in the desert was burned to the ground. An even more convincing demonstration by the devastating "incendiary bats" came when a couple of animals escaped from a careless handler, who thought they "weren't loaded." Fires that resulted consumed most of an auxiliary air base near Carlsbad, New Mexico.

This accident may have dampened the Army's enthusiasm. At any rate, the Navy took over the operation, designating it "Operation X-Ray."

Dr. Adams believes that multitudes of fires could have been set by the bat-transported incendiaries. "We found that bats scattered as much as 20 miles from the point where the bomb [container] opened. Think of thousands of fires breaking out simultaneously over a circle 40 miles in diameter for every bomb dropped. Japan could have been devastated, yet with small loss of life."

But in October, 1944, the project, which is estimated to have cost $2,000,000, was abruptly abandoned. Members of the staff were stunned and mystified by the stop-work order. No reason was ever given, but undoubtedly the order came from high officials who knew that another and more deadly weapon would soon be ready.

... and Bands

Several million birds have been banded in America since Audubon first wrapped silver thread around a phoebe's legs in a little cave in Pennsylvania in 1804. Banding of cave bats in this country was begun in 1932 by Charles Mohr in Pennsylvania and Donald Griffin in New England. Since that time, Mohr calculates that approximately 90,000 bats have been banded by various investigators. "Returns" of banded bats have proved the winged mammals to be remarkably long-lived—up to fourteen or fifteen years—persistent in returning to familiar roosts, and capable of migratory flights of as much as eight hundred miles in one direction.

Banders find the greatest concentration of bats in natural caves and in man-made ones, although sometimes barns and loft buildings are favored homes. In regions without limestone, mines and tunnels often represent the only quarters where bats can find the quiet, darkness, and constant temperature they require for roosting or hibernation.

CHARLES E. MOHR

Anyone who watches the twilight rush of bats from the portals of a cave or their breakneck descent as they hurtle back into the earth is bound to marvel at the accuracy of their navigation. The number of collisions is infinitesimal, and the bats' skill at avoiding obstacles is marvelous to behold.

Although their uncanny ability has been recognized for centuries, it was not until the early days of World War II that the actual method by which bats guide themselves became known. Most scientists believed that networks of nerve endings in the bat's wings were so sensitive to change in air pressure that they somehow warned a bat when an obstacle was at hand.

Sonar or depth sounding, which had been employed at sea since the early 1900's, and radar, which had only recently come into use, were the keys to the mystery. A bio-speleologist and a physicist at Harvard University teamed up to prove that bats use a system which has much in common with both navigating methods.

The investigators, Donald R. Griffin and Robert Galambos, began by confirming experimentally that a blindfolded bat can navigate successfully, but that if its hearing is impaired the animal is unable to avoid other bodies. It soon became apparent that a bat is aware of the position of obstacles by means of sound waves reflected from them as echoes.

How were the sound waves produced? To find out, the experimenters covered the noses and mouths of their bats and released them in the testing laboratory, which was strung with wires. Again the animals were unable to fly with certainty. There was no doubt that the sound was produced by the vocal apparatus of the bats themselves. It was discovered, however, that the direction-finding sounds were not in the audible range of humans. Instead, they were supersonic waves well above the 20,000 a second frequency which marks the upper limit of human hearing. By means of electronic instruments, it was found that bats produce almost continuous pulses in the supersonic range, giving them off at great speed. A single squeak lasts only one two-hundredth of a second.

At rest, a bat's supersonic screech is composed of notes emitted about ten times a second, but on the wing the rate is stepped up to about thirty notes a second, and to fifty or sixty as the animal approaches an obstacle.

Of course, bats do produce sounds audible to human ears: a complaining squeak when they are disturbed during hibernation, and a fairly fast, clicking sound often heard during flight but apparently not associated with the much more rapid supersonic bursts.

Could ordinary vocal cords send out notes with such tremendous energy? It seemed extremely doubtful, because the higher the note and the more rapid the frequency, the greater the amount of energy there must be. Examination of a bat's larynx provided the answer. In most animals the voice box is lightly built of cartilage, but in bats it is comparatively massive and is made of bone. Operated by large and powerful muscles, it is well designed to produce "bat sonar." (The energy used in radar is in the form of electromagnetic or radio waves rather than sound waves.) The echo-location system works

Looking in the east entrance of Kittatinny Tunnel, which housed one of the largest bat colonies in the northeastern states before it was converted into one of the Pennsylvania Turnpike tunnels. *Charles E. Mohr*

so well that failures—collisions—are extremely rare, even in the congested passageways of a major bat cave.

The flight of a great colony of bats has been witnessed, in the United States, by few, even among speleologists, since, except in the Southwest, cave populations rarely exceed 10,000 to 20,000. The largest colony I have found in searching some four hundred caves east of Texas totaled about 120,000 bats, the next largest about 30,000. Fortunately for their safety, several of the most important eastern bat caves are located in Mammoth Cave National Park, where they receive full protection.

In Pennsylvania, where I had banded thousands of bats in an effort to learn how long they live, how regularly they return to their winter hibernating quarters, and where they spend the summer, I tried to visit every cave where bats might stay. In 1938 I learned from Dr. Ralph W. Stone, assistant state geologist, and from geologist Arthur B. Cleaves that the old South Penn Railroad tunnels, abandoned in 1885 when half completed, now sheltered thousands of bats.

Cleaves visited the tunnels while heading survey parties which were studying the feasibility of transforming the series into an all-weather super-highway through the Allegheny Mountains.

"It's just like a *cave*," Cleaves told me as he described the tunnel in the east flank of Kittatinny Mountain. "There are even formations, but they're made of iron carbonate rather than the lime that you would find in a cave. And wait till you hear this! There are bushels of bats! They're everywhere."

That was all I needed. Following the trails cut by the exploration crews, I found it easy to reach the portals. The geologist's description had not been exaggerated. These were man-made caves. Iron deposits along the walls, though less than sixty years old, rivaled in beauty, formations that I had admired in limestone caves. The east tunnel in Kittatinny Mountain extended almost 2,400 feet before it ended at a blank wall. It was dripping with moisture and crowded with dormant bats—thousands of them—more than find shelter in any natural cave in the northeastern states.

In order to hibernate, bats, woodchucks, jumping mice, and other creatures which pass several months in a dormant state, must find a resting place which is dark, quiet, and cool. The abandoned tunnels filled the need nicely. A hundred yards inward from the portals they were darker than night. And the inner sections were unvarying in temperature—55 degrees Fahrenheit—providing an ideal situation, one in which a bat can live in a virtual state of suspended animation. During hibernation the body temperature of an animal drops to that of the surrounding air and its heartbeat and respiration are barely perceptible.

A few hundred feet across the narrow valley from Kittatinny Tunnel, we found the west entrance of Blue Mountain Tunnel, which had become an underground lake. Rock slides from above the portal had created a dam that backed up water for hundreds of yards, to a depth of twelve feet. The geologists had traveled by canoe on this elongated underground lake as they assessed the potential value of the tunnel for highway use.

The west entrance of Blue Mountain Tunnel as it appeared in 1938. *Charles E. Mohr*

We had a few misgivings as we explored the tunnels. How safe were they? Were we taking unreasonable risks in entering them? Dr. Stone, in sending detailed directions for reaching some of the western ones, had written: "West Laurel Tunnel is dangerous. West Ray Tunnel is caved in and very dangerous. Stay away."

As we started exploring the eastern tunnels, we came to feel that perhaps he had exaggerated the dangers. Certainly the tunnels through the eastern ridges of the Alleghenies were remarkably clean. The old railroad ties were in fine condition, with scarcely any fallen rocks. The strata here were nearly vertical, the tunnels like holes cut through a loaf of sliced bread. And in these tunnels, we found the bulk of our bats.

But as we moved westward, making time-consuming "end-runs" around the mountains in order to investigate both portals of each half-finished tunnel, we found fewer bats. And we came to respect the bats' judgment in their selection of hibernating roosts. For the farther west we went, the flatter were the rock layers, and the less stable were the tunnels. In many places the ceilings had collapsed, leaving mounds of rock ten, fifteen, or even twenty feet high. Our inspection was brief, just long enough to be sure there were no bat colonies.

Eventually, when the Pennsylvania Turnpike was completed, one old tunnel was converted into a deep road cut, another was abandoned in favor of a completely new tunnel, and all the others were so reinforced structurally that any danger of further rock falls was eliminated.

"Your bats will have to find a new home," Cleaves told me in March, 1939; "the tunnels are rapidly being converted for the new turnpike." Normally they would be leaving their hibernating quarters shortly after their first spring forays assured them that mosquitoes, moths, and other night-flying insects on which the bats fed, were again abroad in numbers. But in fall, after a season of daytime roosting in barns and hollow trees, they would anticipate the coming of frigid weather by returning to their familiar underground quarters.

The work schedule showed that tunneling operations would be well under way by autumn. There would be bedlam in this long-silent retreat. Before the winter ended, the bats would be dispossessed. Either they would be blasted to death or they would be driven from the tunnels to die of cold or starvation. I could not let that happen without some attempt to save them.

My only hope was to find other hibernating quarters for them. But would they accept a new winter home? Had they become so attached to their tunnel quarters that even if carried away to a safer site they would return to the doomed roost? Since they would be moved in winter, there was a good chance that they might settle down in the new quarters and come back to them again the next fall. We should have to find out.

One December weekend in 1939, Fred Ulmer, co-worker at the Academy of Natural Sciences of Philadelphia, and photographer Jack Ansley traveled with me to the nearest tunnels. We gathered up 150 bats, including a dozen that I had banded the previous winter, and promptly backtracked twenty-five miles to the outskirts of Carlisle. Here we found Conodoguinet Cave on the banks of the creek of the same Indian name, meaning "for a long way nothing but bends."

The bats we had brought from the tunnel were now wide awake, so after we had banded each one we tossed it into the air. Most of the bats flew directly into the cave and disappeared, but at least two dozen of them circled over the creek once or twice and then headed west, in a beeline for the tunnel. Would the rest stay in the cave? We wondered. Workers were surveying a section of the turnpike directly over Conodoguinet Cave when we returned five weeks later to look for our banded bats. Fred Ulmer was the first to spot one, hanging just inside the entrance of the cave and so covered with sparkling dew that it looked white. Its number was 1736—one of the tunnel bats.

"Here's another, but it's dead," called Jack Ansley. He had found a bat on the muddy cave floor. It was number 1849, another of our tunnel bats. "Here's another dead one," shouted Fred from on ahead. In a few minutes we counted ten dead banded bats. In each case they had been trampled into the mud. The tracks were human footprints, not animal. Very likely the bats had been knocked from

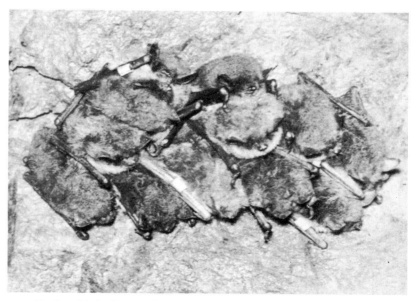

Twelve Little Brown Bats, *Myotis l. lucifugus*, hibernating in a Pennsylvania cave. Four arm bands are clearly visible. *Charles E. Mohr*

their roosts by superstitious or ignorant persons who had ventured into the cave.

In an obscure crevice, I found two live ones, 1847 and 1848. We saw others undisturbed, bats beyond the reach or sight of marauding visitors.

This cave obviously was too well known and offered too few hiding places. It would be foolish to bring any more bats here. But perhaps they could be transferred successfully if we picked a cave more remote from civilization. I had just the spot in mind—Aitkin Cave, seventy-five miles to the north in the cavernous limestone country between Lewistown and State College. I had no doubt that the cave would be suitable for bats. It already had the largest population in the state.

We had come prepared to move thousands of bats if our trial were successful. Since it seemed to have been, we went on to the tunnels, and while helmeted tunnel workers stared in open-mouthed astonishment we raked hundreds of bats into cages, buckets, and finally even into burlap bags.

We had arrived in the nick of time. Tunneling operations were crowding the bats into a steadily diminishing section of the old tunnel. Fully 5,000 bats hung in scattered clusters within a few hundred feet of the end of the workings, quite undisturbed by the terrific din set up by the drilling equipment and the exhaust fans. Water, sprayed on the heading to wash down the dangerous silica dust produced by the drilling, fogged our spectacles and camera lenses but didn't seem to bother the bats. Dormant when we loosened their grip on the rough walls, the bats soon were chattering excitedly and crawling over the wire screening of our cages.

Grimy, big-muscled miners gathered around, fascinated by the winged creatures which most of them had feared since childhood. Their unprintable exclamations betrayed the general ignorance of, and hostility toward, bats. Good riddance, they said, in effect. Some of the workmen, though, wanted to know more about them and what we were going to do with them.

We planned to band several thousand before we released them in Aitkin Cave, we said. We hoped the bats would adopt the cave and not return next year to the inhospitable tunnels. Most of the men were still shaking their heads as we packed up our nets and crates and headed for the car.

A group of about 350 Little Brown Bats hibernating in Aitkin Cave in central Pennsylvania. *Charles E. Mohr*

The weather turned against us before we reached Aitkin Cave. Freezing rain made driving hazardous and soaked us thoroughly on the quarter-mile trips we had to make between the car and the cave with our equipment and cages. For hours we sat, cramped and wet, in the ice-filled entrance of Aitkin Cave, painstakingly attaching the tiny aluminum bands and recording the species and sex of the bat and the number of each band. By the time we had banded a few hundred and examined several hundred more, we were too exhausted and cold to continue.

Reluctantly, we released thousands of bats unbanded, satisfied that at least we had given the useful little animals a chance for survival. Carrying our empty cages, we trudged back through the rain to our car and set out for Philadelphia. That was the last time we ever saw any of the tunnel bats.

That spring and summer the turnpike was being pushed toward an amazingly swift completion. By fall the tunnels were in operation.

Many times Fred Ulmer and I talked about the bats. Where were they vacationing? Would they go back to Aitkin Cave? Or would they become victims of well-established habits and seek their familiar haunts in the tunnels of southern Pennsylvania?

When winter finally had settled in, we drove upstate, got into our caving outfits, and began the search for the tunnel bats. Aitkin Cave was first on our list. There were great mats of bats covering the smooth, arched walls, filling in the spaces between pincushions of stalactites, cramming bowl-like pockets in the ceiling.

Not a single banded bat could be found. Perhaps they were here, and we had just failed to find them among the scattered clusters. Carefully we calculated the size of the wintering population. Was there an increase over previous counts? No, the total count was the same as it had been in other years. They must be elsewhere—perhaps in the tunnels.

These had been opened to traffic in October. Speeding across the new turnpike to the first mountain ridge, we found a well-lighted, concrete-lined tunnel. In a minute we were through Blue Mountain. Just beyond, we pulled off the road in front of the portal house at Kittatinny Tunnel.

"You fellows should have been here this fall," we were told when we had identified ourselves. "'For six weeks there were bats flying in and out of the tunnels. Hundreds of them every night until cold weather set in."

What had become of them now, we wanted to know. The bats had gradually given up their futile search for hibernating quarters in the well-lighted, smooth-walled tunnels, and had disappeared. Had they been able to find shelter in natural caves within easy flying distance of the tunnel?

Since there were quite a few small caves just to the south, we continued our search there. Around Shippensburg we looked for bats in Carnegie Cave, the Cleversburg Sink, and Hershey and Kelly Farm Caves. Only a dozen to a score of bats in each—*no* bands. Where else might they be? Carlisle?

We headed back east toward our starting point, making futile searches of Craighead Cave and Boiling Springs Cave. Finally we returned to Conodoguinet, the little cave beneath the turnpike, the one with almost as many bends in its passage as in the quiet stream that flowed beside the entrance.

But even in Conodoguinet Cave there were none of our tunnel bats. Where had they gone then? Some may have found shelter from winter's paralyzing cold in crevices too small for man to enter. Most of them probably perished. For years I held to the hope that some sharp-eyed spelunker would spot a few banded individuals. But year after year passed without a report, until I was forced to conclude that most of them died the winter the Pennsylvania Turnpike tunnels were opened.

Howard N. Sloane *(left)* and Charles E. Mohr relaxing in a Pennsylvania cave, while on one of their numerous expeditions. *Bruce C. Sloane*

The One Who Cries

The story of Western Hemisphere caves would be incomplete without reference to the most celebrated cave in South America—Guacharo Cave. The first cave in America of which there is written record, the home of the remarkable cave-dwelling guacharo, it has been visited by a succession of prominent naturalists, from 1799 to the present.

Wholesale slaughter of the cave-nesting birds, an annual ritual for untold centuries, has long posed a problem for conservationists. World-wide interest in the cave and its unique inhabitants has finally brought protection for both.

Some of the credit for present-day interest can be traced to the enthusiasm and energy of one of Venezuela's prominent speleologists Eugenio de Bellard Pietri, who has generously supplied some of the information on which this story is based.

HOWARD N. SLOANE

Its scientific name is *Steatornis caripensis,* but in Venezuela it is called *guacharo*—"one who cries and laments." There are only three or four thousand of these strange birds, and all of them live in northern South America. Colombia, Ecuador, Peru, and British Guiana have a few, but the greatest concentrations are in Venezuela and on the Island of Trinidad, ten miles off the coast.

Guacharo Cave, the home of the largest single colony, is situated at an elevation of 3,000 feet near the village of Caripe, Venezuela, 240 miles east of Caracas. Here live more than a thousand of the total guacharo population, but the bird is rare even in this country. Speleologist Eugenio de Bellard Pietri searched in more than forty Venezuelan caves, yet found this bird only in Guacharo Cave.

This cave is probably the first in the Americas of which we have written record. In 1678 Father Tauste, a Franciscan-Capuchine

The huge entrance to Guacharo Cave, eighty-five feet wide and seventy-five feet high. The Indians often held religious ceremonies in the cave entrance before proceeding with the "oil-harvest." *Werner Cohnitz*

missionary, described the cave and the astonishing "oil-harvest" through which the guacharo became known as the "oil-bird."

During midsummer, when young guacharo chicks are most abundant, the Chaima Indians, the aboriginal inhabitants of this region, used to celebrate an annual festival. Hundreds of natives overcame their superstitious fears of caves and their reluctance to get within earshot of the horrible screeches of the guacharo. Carrying extremely long poles, they invaded the cave and knocked down every nest they could reach. The guacharos normally nest on ledges in the uppermost thirty feet of the cave, and it is rare for a bird, even in flight, to descend within thirty feet of the cave floor. As the round clay nests fell, other natives waited ready to collect the young squabs. They ripped them open on the spot and removed the peculiar fatty layer which stretches between the young birds' legs from the abdomen to the tail. This prized mattress of fat disappears as the birds mature.

Leaving the cave floor littered with dead birds, the natives took the fatty portions to the cave entrance where palm-leaf huts had been erected, each provided with a brushwood fire and clay pots in which to melt the fat. When rendered, the fat was called "butter" or "oil of the guacharos." Half liquid, transparent, and odorless, this "butter" is so pure that it can be kept for a year without becoming rancid. It was formerly used in cooking by natives and missions alike.

The wanton slaughter, which may have been practiced for generations before Father Tauste described it in the seventeenth century, continued until 1949. Had it not been for the fears of the inhabitants which kept them out of the cave except during the "oil-harvest," and for the difficulties in reaching the lofty nesting perches, the birds almost certainly would have been exterminated. Hundreds, sometimes thousands, of young birds were destroyed each year.

Finally on July 15, 1949, through the efforts of speleologists and other scientists and conservationists, the Venezuelan government decreed the guacharo area a national preserve called Alexander von Humboldt National Park. Park service officials now stationed there have effectively stopped the killing of the guacharos.

Baron von Humboldt, for whom the park is named, explored Guacharo Cave in 1799, and his description and observation of the cave and the guacharo constitute the most complete account of both until recent times. Since Humboldt's period few white men, except scientists interested in learning and recording the details of the guacharo's life, have seen these birds. Within the last few years new interest has been aroused, and visits have been made by ornithologists William H. Phelps and William H. Phelps, Jr.; Dr. Eduardo Rohl, Venezuelan scientist and naturalist; William Beebe; speleologist Eugenio de Bellard Pietri; and Donald R. Griffin.

What so fascinated these men that some of them traveled many miles and through many hardships just to see a bird in a cave?

Though its body is only about the size of a chicken, its wingspread is nearly four feet—as great as that of many of the larger hawks. The plumage of both sexes is a reddish-brown or chestnut color, slightly greyish underneath. The wings, tail, and other portions are marked with small white spots edged with black. Some striation or alternate light and dark coloring occurs throughout the rest of the feathers. The beak is hooked and strong and provided with a double tooth. The legs are pink and sturdy but without talons, for the gua-

Eugenio de Bellard Pietri holding a young guacharo in Guacharo Cave. The adult bird has a wingspread well over three feet. *Roberto Contreras*

charo is not a bird of prey. It is related to the nighthawk and the whippoorwill, nocturnal insect-eating birds of North America.

Ivan T. Sanderson, well-known explorer, lecturer, and collector of rare animals, describes the guacharo in his book *Caribbean Treasure*: "It was a beautiful creature in its rather unearthly way, rich brown in colour with large white spots, great enormous . . . staring eyes set in a low, perpetually angry face that could open right across into a gaping mouth surrounded by bristles and armed with a beak not unlike that of an eagle or hawk. Its tail was a perfect fan; its wings slender and tapering. It exuded an overpowering, musky odour that was exceedingly unpleasant, extraordinarily penetrating, and remarkably lasting."

Several writers have commented on the birds' blue eyes, accustomed only to the total darkness of the cave and the faint luminosity of the moon and the stars. The guacharo leaves its cave only at night.

Each evening between six and six-thirty all adult guacharos set out to forage for food. A vanguard of birds stages a preliminary flight, returning to give the flock the "all-clear" signal. Compact groups of about a hundred birds are followed rapidly by other groups until not a single adult remains.

On their return, always before dawn, they re-enter in groups and disperse in the cave, flying to their nests to feed their young with the fruits and nuts they have stored in their crops. Certain seeds found in the dropping in the cave are from species of trees which botanists have not found within 170 miles of Guacharo Cave. For this reason it is believed that the birds make flights of that length in a single night.

In 1949 Ivan Sanderson and the author secured permission from the Venezuelan government to visit Guacharo Cave and bring six birds back to New York City, but transportation difficulties, and the uncertainty of securing an adequate and suitable food supply for them, made us abandon the project. Sanderson has developed a system for feeding guacharos by hanging fruits in specially built dark corners of their cages. In this way he has kept them in captivity for limited periods, but all attempts to keep guacharos in permanent captivity have been unsuccessful.

Guacharos are noted for their ability to create an unbelievable volume of noise. In *Caribbean Treasure* Sanderson described his reaction to the cries of the *diablotín,* as the guacharo is called in Trinidad:

"There we stood in the brilliant world of life, surrounded by all the colour and full-bloodedness of a tropical forest, and peered into a cold dark void that was nevertheless alive with unseen things that betrayed their hidden presence by some sense other than sound.

"We descended the slope and stood on the brink just below the edge of the great arch. The silence before us was absolute. Then once again came that obnoxious noise, but this time a hundred times louder, more bathroom-like and insufferable. It was the signal for an outburst that did not abate for several hours.

"Out of the darkness within the earth the harsh screams of a thousand devils were let loose, and myriads—I am not exaggerating—of winged things set off from one hidden place to another, each mass making a noise like an airplane as it got under way. Things streaked across and in and out of the mouth of the cave, little things that twisted and whirled in the half-light, while high up big things flapped for a moment and then were swallowed once again by the gloom. The din was terrific. Above everything else rose those harsh rattling screams, the most nerve-racking noise I had ever heard and one to which I could never become accustomed."

Humboldt wrote: "It would be difficult to form an idea of the frightful noise which thousands of these birds produce in the dark portions of the cavern . . . their shrill and piercing cries strike upon the rocky vaults, and are repeated by the subterranean echoes."

When Donald R. Griffin read Humboldt's account, he wondered whether guacharos have an echo system of navigation comparable to the one he and Robert Galambos demonstrated in bats. In 1953 he visited Guacharo Cave to determine the degree of darkness in which the birds could fly and to discover, if possible, whether they, like bats, use a "sonar" system to avoid obstacles.

In Guacharo Hall, while birds flew noisily overhead, Dr. Griffin exposed photographic film for nine minutes. When it was developed, the film showed no evidence of having been exposed to light. There was no question about it—the guacharos fly in total darkness!

How did they navigate? By means of loud cries? Or did they, like bats, produce notes in the supersonic range, inaudible to human ears?

Using amplifiers, electronic filters, an oscillograph, a tape recorder, cameras, and other equipment, Griffin found that in flight the birds issued a stream of sharp clicks, each two thousandths of a second in duration. Unlike the signals of bats, however, these sounds were well within human auditory range. But were these clicks for flying purposes?

Capturing several birds, Griffin experimented by stuffing their ear canals. Thus hampered, the birds were unable to fly in the dark, and banged into walls and obstructions. When the ear plugs were removed, they could again fly properly. In light they flew without difficulty, with or without ear plugs. They did not need "sonar" so long as they could see where they were going.

Dr. Griffin logically calls this method of locating objects by means of echoes, "echo-location," and compares it to the ability of blind persons to approximate their position by echoes from the taps of a cane or other noises.

Guacharo Cave is estimated to be about two miles long. Its huge entrance is eighty-five feet wide and seventy-five feet high. It is reasonably level in the beginning and consists of two major divisions; the front and better known portion, and a rear section called Hall of the Wind. The Guacharo River, only one to two feet deep but fifteen to twenty feet wide, traverses the entire first part of the cave. This section is divided into three successive galleries—Guacharo, or Humboldt's, Hall, 2,685 feet long; the Hall of Silence, 620 feet long; and the Precious Hall, 370 feet in length.

Guacharo Hall is a gallery which averages seventy feet in width and has a ceiling height ranging from eighty to one hundred feet. It contains massive stalactites but no stalagmites. Daylight reaches about 450 feet into the cave, to a point where the passage changes direction and the light cannot penetrate. Shortly beyond, an explorer can faintly hear the sounds of the birds from deep in the cave. The

The entrance room to Humboldt's Hall. The guacharos are found at the back end of this room, almost half a mile beyond this point. *Werner Cohnitz*

Germinating seeds from the droppings of guacharos produce dwarfed, yellowish plants which soon die from lack of sunlight. *Werner Cohnitz*

noise increases in intensity as one approaches the main roosting area high at the rear of Guacharo Hall, a half-mile from the entrance.

The Hall contains fantastic amounts of droppings, or guano, accumulated through the centuries. Many of the seeds and kernels from fruits eaten by the birds germinate in the abundant guano, and little forests grow wherever the birds roost overhead. Of course, the absence of sunlight precludes development of chlorophyll, and all the vegetation is white or yellowish and stunted. Most of the germinating seeds are from the *palmito* tree, which in sunlight attains a height of sixty feet. In the cave these miniature white palms grow only a foot and a half high.

The next room in the cave, the Hall of Silence (so called because it echoes only faintly with the shrill screams of the birds in the previous room) is narrower, and the ceiling height averages only about twenty-five feet.

The Guacharo River disappears in the Precious Hall. This chamber, more irregular than the others, has a ceiling height nearly equal to that of Guacharo Hall but varies in width, averaging about forty-five feet. It contains beautiful calcium carbonate formations. To the right, at the rear of the room, a twenty-foot pit connects with the lesser known parts of the cave. There is danger here for the inexperienced explorer. The cave is no longer level, and exploring it means crawling on narrow ledges along the brink of pits, squeezing through ugly crevices, climbing steep boulders, and traversing two passes, one by diving. This is Wind Pass, where a constant rush of air flows toward the cave entrance. Getting through Wind Pass necessitates diving sideways in frigid water four to five feet deep at a point two feet below the surface, and maneuvering for a distance of twelve feet.

The balance of the cave is even larger than the known portion and according to Pietri presents an amazing display of "blood-red crystalline stalactites, stalagmites, helictites, spaghetti-like formations and columns—all unspeakably beautiful. . . . Every square yard of the floor differed from the preceding one with lovely lace formations."

The size and weirdness of the cave, the disturbing noises issuing from its depths, and the nocturnal flight of the birds all helped to create mystical ideas in the minds of the natives. It was only during the "oil-harvest" (and in the company of many others) that the Indians could be induced to penetrate beyond the point reached by daylight.

Humboldt wrote: " 'to go and join the guacharos' is with the natives, a phrase signifying to rejoin their fathers, to die. The magicians and the poisoners perform their nocturnal tricks at the entrance of the cavern, to conjure the chief of the evil spirits."

"The natives," said Humboldt, "connect mystic ideas with this cave, inhabited by nocturnal birds; they believe that the souls of their ancestors sojourn in the deep recesses of the cavern. 'Man,' say they, 'should avoid places which are enlightened neither by the sun (xis), nor by the moon (nuna).' "

The Idol of the Cave

Curator of Archaeology and Ethnology at the Southwest Museum, Los Angeles, M. R. Harrington is well known for his excavations and book Gypsum Cave, Nevada. *He was co-author, with L. L. Loud, of* Lovelock Cave. *Harrington has taken an active part in the National Speleological Society's archaeological projects and was named Honorary Member for 1950. He is a Fellow of the American Association for the Advancement of Science and a Research Associate of the Carnegie Institution of Washington.*

While associated with the American Museum of Natural History in 1900, he excavated Leatherman's Cave, near Greenwich, Connecticut, his first speleological enterprise. Later he studied Ozark Cave remains. In Cuba he carried on important investigations for the Museum of the American Indian, Heye Foundation, which produced a two-volume work, Cuba before Columbus. *It was during that expedition that he had the experiences described in this chapter.*

M. R. HARRINGTON

"People tell me, Señor, that you have discovered a cave at La Patana in which you found hairy spiders knee-deep. Is that a fact?"

I had to inform the gentleman that he should not trust back-country gossip, especially the brand current about Cape Maisí. The term he used, *arañas peludas,* in Cuba refers to tarantulas. We did not find tarantulas knee-deep in the cave, but what we did encounter was almost as extraordinary, and unexpected.

My first intimation that such a cave existed came some time before, while my archaeological expedition, sent out by the Museum of the American Indian, Heye Foundation, was quartered at the little village of Jauco on the south coast of Cuba, some distance west of Maisí. Well I remember the evening when a gaunt young Cuban, his once-white suit smeared with the red clay of the hills, arrived from La Patana, bearing a mysterious bundle swathed in cotton sheeting.

Opened, it disclosed a number of little human and animal figures made of pottery, which seemed to be handles broken from bowls, and several polished-stone hatchet blades that were very well made. Finally, there was a beautifully finished wooden platter with carved decoration and a grotesque head on one end for a handle. It was of ancient Indian make but unlike anything I had ever seen. We were amazed, because our own digging thus far in Cuba had produced only very crude implements of shell and stone.

"From a cave near my home," the young man explained. I knew the platter must have come from a cave, and a very dry one at that, or it never could have survived the centuries. "My family, the Mosqueras, own the place."

"There are other caves at La Patana," he went on, "some very dark and very large, that we fear to enter. Strange sounds are heard within them, and the air has a sickening smell. In one, they say, there is a *zemí,* an idol of stone worshiped by the Indians long ago; and the grandfathers tell us that the monster serpent of Maisí, the *Culebra Serpiente,* dwells somewhere within them."

I had heard yarns of the "Snake Serpent," and I might have thought they contained a grain of truth if the storytellers had not always insisted, the creature "crows like a rooster." That was too much to swallow. The story of the idol might be a fairytale, too; but if it really existed—and we could find it—it would be a prize for the museum. At any rate, the caves of La Patana should be explored.

Our visitor explained that the name La Patana referred to a wooded terrace—he called it a mesa—on the south flank of the great limestone plateau that forms the eastern tip of Cuba. There were five or six such terraces, he said, forming giant steps from the flat summit down to the sea. La Patana, named from the Patana cactus, lay about halfway down.

Finally, our work at Jauco was finished. By this time we had discovered why the first specimens our excavations uncovered were so much cruder than the fine articles from La Patana: They had been made by a different people—an earlier, more primitive tribe, who had preceded the more advanced Taino Indians found by Columbus in eastern Cuba.

The early people, whom we called "Ciboney," apparently raised no crops but lived entirely on the natural products of the forest and the ocean—mainly the former, to judge by the great number of land-snail shells and land-crab claws they left behind in their ash dumps. In place of pottery, they made bowls and dippers of large conch shells, and conch shell instead of stone was their material for hatchet blades and gouges. How long they lived in Cuba we have no way of telling; but the bones of a large extinct animal, the ground sloth (*Megalocnus*), are sometimes found in Ciboney cave deposits, and this suggests a long time.

After them came the Tainos, an agricultural people, growers especially of yucca, or cassava, and of corn. In fact, our word "maize," applied to Indian corn, comes from the Taino language. They were expert potters and loved to decorate their work with little figures of men and animals, modeled in the round. Their ax heads were of stone, often beautifully shaped and highly polished. Their wood carving was excellent, and they were skilled in making ornaments and little charm statuettes in shell, bone, and stone. The things brought to us from La Patana were undoubtedly Taino products.

So it was with high expectations that we moved our headquarters eastward to the Finca Sitges, an extensive coffee and banana plantation on top of the plateau, owned by the hospitable Don Antonio Rey. By this time our party had received a welcome addition—a representative of the Cuban government, Dr. Victor J. Rodríguez, who was a zoologist from the University of Havana. Of course, La Patana was one of our first projects.

Following a narrow trail and burdened with packs of food, hammocks, and blankets, we climbed down over limestone cliffs and crossed stretches of orchid-hung tropical forest. Finally, once more descending, we sighted the palm-thatched roofs, beehives, and broad-leaved plantain trees that told of human habitation. In a few minutes we were talking with the elder Mosquera, father of the young man who had visited us. He assured us that we could explore the caves to our heart's content. They used only one, which contained the spring from which they drew their water.

I asked him why they had no trail for saddle or pack animals. "It is not necessary," he said. "We raise most of our own food here. If we need anything from the store, we trade honey and beeswax for it, and that can be carried easily on foot. If money is needed, the boys go up and work on the *fincas*."

He detailed one of the *niños*—his son Cecilio—to show us around. Depositing our packs and taking only trowels, camera, carbide lights, and a lantern, we set out.

First he led us along a well-beaten trail to the Cueva del Agua, source of the water supply for the Mosqueras. Passing through the high-arched entrance of the cave, where stalagmites stood, looking from a distance like groups of statuary, we soon reached the crystal-clear spring. It lay at the very back, yet was still in plain daylight. We searched and searched but found little trace of ancient Indians, except for a few bits of pottery at the very entrance. Then we followed Cecilio to a passageway leading back into the mountain, which we had not noticed before.

"Where it leads, I do not know," he said. "We have never dared to follow it. But you have good lights; with you I am not afraid."

With some trepidation we set out. The passage dipped slightly downward, turning to the west. We had just begun to feel a little

The cave at La Patana has many native names—Horror Cave and Cockroach Cave are two favorites. Here, the expedition gathered in the entrance room for a picture taken by Dr. Harrington. *Museum of the American Indian, Heye Foundation*

more at ease, when suddenly it ended—in a great black gulf. We saw this just in time.

We approached the edge cautiously and shone our lights into it. Apparently the chasm had neither end, nor top, nor bottom. It was Rodríguez who thought of dropping a loose stone into the abyss. After an interval, we heard it ring on the bottom. "It does have a bottom after all," he remarked with a wry grin.

As we stood, a few bats fluttered out of the darkness, wheeled about our heads, and disappeared again. Opening his kit, Rodríguez produced a butterfly net, hoping to catch new species of bats. We turned our lights down so as not to scare the creatures, and it was while Rodríguez was trying to net one that we first noticed the peculiar acrid odor wafting up from the depths.

And then we heard it—a strange, unearthly, roaring sound, which grew louder and louder. We all stepped hastily back from the brink. Gaspar Leiba, one of my helpers, half drew his machete. Bats were forgotten.

We looked at each other in amazement. Now the sound gradually dwindled away until it was nothing but a low murmur. I found my voice first.

"What in heaven's name is that? I've explored a lot of caves, some right here in Cuba, but I never heard anything like it." Of course, it couldn't be the *Culebra Serpiente,* or could it? The idea was not pleasant. Also, I am ashamed to say, I thought of the ground sloth, a large extinct animal whose bones we had unearthed in another cavern. Could monster serpents or ground sloths roar?

"I can't understand it," Rodríguez said. "It couldn't be a waterfall deep underground, because the noise is not continuous. It might be surf in a sea cave, except that we are too high above the ocean and too far back.... Listen!" The roar was beginning again.

This time we stood our ground, peering over the brink, but no monsters appeared. All was black darkness as before, but we noticed for the first time a faint glimmer of daylight, distant and far below.

"See that light?" Cecilio asked. "I think that is another entrance, and perhaps I can lead you to it. I went in one day when I was a boy, but something roared at me and I came out flying."

He led us out of the cave, down a cliff, and around a projecting point of the mountain. There, along its base, stretched a great hole, open to the sky and perhaps fifteen feet deep. Fortunately, a large *jaguey* tree grew in the near end, and this we used as a ladder to clamber down. There were two cave mouths running from the hole back into the mountain.

Cecilio picked the easternmost, and in a few minutes we found ourselves in a huge dark chamber. The smell was stronger than before, and again we heard mysterious murmurs and roars. The sounds seemed to come from a passageway blocked with fallen limestone on the west side of the cavern. However, to be sure we were on the right track, I sent Gaspar with a lantern to retrace our steps. Sure enough, after an interval we saw his light emerge high above us on the opposite wall.

After his return we tried to enter the blocked passage but could not get through. Then Cecilio suggested that the other cave mouth might tap the same passage on the other side of the rockfall. He was right.

The smell here was much stronger, and the roaring, when it came, louder still. And what a sight lay before us! The floor of the passage, as far as we could see, was carpeted with living, moving cockroaches. Rodríguez, who wore no leggings, kept his legs moving to prevent the repulsive creatures from crawling up his trousers. And every now and then we saw a big, shiny centipede winding solemnly along. One, I noticed, had a cockroach in its mouth.

We started walking along the passage, crunching cockroaches at every step. Now the air grew hotter, the roars were louder, the smell was almost overpowering. Bats fluttered overhead. Soon there appeared an added horror; the walls were now decorated with huge, black, spider-like creatures. We later identified them as a species of tailless whip scorpion, with one pair of legs lengthened for use as feelers in the darkness; but at the time, they looked like big black spiders to us.

As we went on, it grew still hotter and more oppressive; the lights flickered, and the kerosene lantern went out. We might have continued, even so, but for one thing. I spied a mound of fighting, struggling cockroaches and kicked them aside to see what they were after. It was a baby bat, already half picked to a skeleton, although the poor creature was still moving a little.

What would happen to *me* if I were overcome by the bad air in that awful place? Already I felt a little dizzy. The same idea must have struck us all at the same moment, for we turned and beat a retreat without a word.

Climbing back up the cliff, we drew a couple of buckets of water from the spring and washed up. Then, after retrieving our packs, we slung our hammocks for the night from trees near the Cueva del Agua. The Mosqueras had invited us to stay with them, but we politely refused. We were too tired and frustrated and did not even enjoy our supper. It did not improve my own state of mind to realize that we had not seen a trace of ancient Indians since we picked up those bits of pottery. Where was the Idol of Cape Maisí?

As night approached, we noticed a black cloud of bats emerging from the cave mouth, setting forth on their nightly search for food. To check on what might be happening at the last cave we had visited, we made our way down the cliff again. What a spectacle! Here they were pouring out in swarms. It seemed to Rodríguez that there were different species and that the bats emerged in groups, those of the same kind flying together.

After about an hour, we returned to our hammocks; but the bats were still coming—fluttering black blotches against a fading sky.

Rodríguez came over and sat on the ground beside my hammock. "You know," he said, "I think the heat and the bad air in the cave are due mainly to bats. As you saw this evening, there are an enormous number of them. While they are out hunting food at night, the cave should cool off, and maybe they sweep in a little fresh air when they return at dawn. If we try again early in the morning, perhaps we can get somewhere."

It was so agreed. Rising at daybreak, we gulped down our coffee and hardtack. As a matter of precaution, I belted on my .38. I can still picture the little force: Rodríguez, brown-eyed, pale-faced, rather frail of body but determined in spirit, and armed with a collecting pistol; Juan Gauche, my *ayudante,* of pure Spanish ancestry, hook-nosed, thin-faced, and as blue-eyed as they come. He carried a machete, as did the other two, who showed plainly their Cuban Indian blood—short but well-built Gaspar Leiba and lanky Cecilio Mosquera. As an extra piece of equipment, Gaspar carried a large canvas bag, to use if we found something worth collecting.

Soon we were again treading the cockroach corridor. Rodríguez was right. The air was distinctly cooler, much more breathable; the lights burned steadily. Even the odor seemed less pungent. But the roaring—that was as loud and frightening as before. We passed the place where we had turned back previously, and pressed on and on. Suddenly an unusually loud roar halted us in our tracks. I confess I grabbed for my gun. All at once the air seemed hotter again, more stifling. The temptation to turn back was strong.

But the sound died away, and we forced ourselves to go on. Soon we were standing on the threshold of another great chamber. We shone our lights about. Wherever we looked, the high vaulted ceiling was covered with clinging bats! Besides these thousands there were hundreds more hovering about, looking for a place to alight.

"Listen!" Rodríguez whispered.

A murmuring sound was plainly audible. One bat is silent enough, but hundreds of them, in that confined space. . . . There was no doubt about it; the murmur was the sound of their wings, reverberating from the vaulted roof.

"I'm going to try something," he continued. "Watch this!"

"*Hola!*" he shouted.

Instantly every bat in the cave took wing; the murmur rose to a roar.

We stood there, dumbfounded. Gradually the sound decreased as the frightened creatures returned to their places.

We had solved our major problem—and a minor problem as a result; the infernal smell in the cave. It was "the concentrated perfume of bats," as Rodríguez phrased it. The floor of the cave was composed mostly of bat droppings, the accumulation of nobody knows how many years.

A more difficult question was the cockroaches. What did they live on? It seemed improbable that there were enough fallen baby bats or dead adults to keep them alive. It seemed likely that they found pieces of bugs in the droppings of the insect-eating bats, and Rodríguez offered a further possibility. Scooping up a handful of bat guano from the floor, he said:

"Do you see all these rough-looking little sticks mixed in with the droppings? I think they are *jubo* berry cores." (These berries grow on trees and have fibrous cores something like American mulberries.) "Many of the bats are fruit eaters," he continued. "Perhaps they bring back berries to chew on; the cores drop to the cave floor, and the little *cucarachas* suck the remaining juice." In any event, the cores got into the cave somehow, and the roaches must have enjoyed them.

On our way back along the cockroach corridor, I walked close to the right-hand wall, looking into holes and cracks where the ancient Indians might have stowed away something of interest, as indeed they had hidden the wooden platter. Coming to a larger crevice, I peered in and found myself face to face with a big snake neatly coiled in the opening, its nose not ten inches from mine! Jumping back, I called Rodríguez. He investigated cautiously.

"It's what we call a *majá*," he said, "a relative of the boa constrictor. I'm going to try to catch it for the zoo in the Parque Colón at Havana. Don't worry, it's not poisonous. In fact, we have no poisonous snakes in Cuba."

He waved his left hand in front of the hole while Juan shone his light upon it; he kept his right hand at one side, out of sight. I saw the snake raise itself, watch for a moment, then stick its head out of the crevice to see at close range what was going on.

Instantly Rodríguez grabbed it by the neck, and the struggle began. After some minutes, seeing that my friend could not make it alone, I took hold also. How long we pulled and sweated, I don't know. Sometimes we had the snake nearly out of the hole; then, in a panicky burst of strength, it nearly got away from us again.

Finally, Rodríguez had an inspiration. Calling Gaspar, who was still carrying the canvas sack, we slipped the snake's head inside it, then turned the reptile loose. Apparently content to find peace and quiet in the sack after so much commotion, the reptile let go with its tail and of its own accord slid its eight feet of length into the receptacle. Rodríguez tied up the sack, and that was that.

As we headed again for the entrance, a loud roar from behind startled us. We knew now what it was, yet instinctively quickened our pace. In fact, we were almost running when we emerged into daylight.

Once in the cool cave mouth, we sat down to recuperate. Gaspar, gingerly depositing the snake bag on a smooth spot, announced, "Now, with permission of the señores, I shall bathe myself." He shed his cotton coat and trousers, kicked off his *alpargatas* or sandals, and stepped into a pool of water on the east side of the opening.

The rest of us, equally sweaty and dirty but less ambitious, watched him with envy. Suddenly I noticed, on the cave wall beyond the pool, something unusual. The whole surface was covered with ancient Indian carvings, known to archaeologists as "petroglyphs." They were in plain daylight, although partially hidden by an encrustation.

Now, with renewed interest, I looked about the cave mouth for further traces of ancient Indians. Not fifteen feet away, I was astonished to find that a large stalagmite we had passed several times on our way into the cave was crudely carved to represent a human figure! It not only had a plainly marked face but indications of a body carved upon it. It was the *zemí*, Idol of Cape Maisí.

The stalagmite, about four feet high, stood some 50 feet back from the shelter line of the cave mouth, in plain if subdued sunlight. Most of the carving was on the surface of the stalagmite that faced east. Here we found a face made by pecked-in grooves clearly representing the mouth, nose, and eyes. Other grooves suggested limbs and male genitals, and still another encircled the forehead like a headband. This image was so placed by nature that at a certain time in the morning, at least during our stay in June and July,

Petroglyphs over a water pool in Cueva del Agua. *M. R. Harrington.*
Museum of the American Indian, Heye Foundation

The Idol of the Cave

The *zemí*, the Idol of Cape Maisí, as it was originally found. In addition to the face, indications of a body were carved on this four-foot-tall stalagmite. *M. R. Harrington. Museum of the American Indian, Heye Foundation*

a shaft of sunlight striking through a crevice fell full upon the face of the figure for a few minutes.

The north, south, and west surfaces also bore rude faces indicated by shallow depressions, but these were less detailed than the east face. In order to photograph the north face, for example, we had to whiten the grooves with cassava starch.

Who made the idol? We consider it a product of the Taino Indians, who had carved the beautiful wooden platter, but the workmanship is crude for such skillful craftsmen. Perhaps it dates back to the days of the Ciboney tribe.

It seems probable that the mysteries of the hot, dark, subterranean chambers—the roaring sound, the millions of cockroaches, and thousands of bats—existed in Indian days as well as now. If so, these awe-inspiring phenomena may well have caused the selection of this particular cave as a special spot for "cavern worship," known to have existed also among the related Indians of Haiti.

But alas for my dream of finding a real prize for the museum! I had pictured a statue of stone or wood that could be picked up and carried. The idol, being a stalagmite, was anchored to the rock bottom of the cave! Moreover, it must weigh four hundred or five hundred pounds—far too heavy for any pack animal to carry, even if we could pry it loose. And if we succeeded in finding a strong enough pack animal, there was still no trail from La Patana to the outside world.

All I could do was to photograph the idol and the carvings; and when we returned to Finca Sitges, I mentioned them in my routine report mailed to the museum. Regretfully, I thought that was the end of the matter.

But I was wrong. Coming in from a trip one day, I found a cable from Dr. Heye, the director. *"Get that idol"* was all that it said.

Plainly, a trail was the first necessity, so we went scouting and found that to build one up the cliffs was out of the question; it would take too long. Our only hope would be to follow the terrace of La Patana eastward and connect with the old trail leading from the top of the plateau to the lighthouse at Cape Maisí. That would take only days, not weeks. So we got busy rolling rocks aside and where necessary cutting a path through tropical growth with machetes. In one place, where the limestone underfoot was full of holes and sharp points, we had the unexpected job of filling in these *dientes de perro* ("teeth of dog"), as they were called by the local *cubanos*. When this job was completed, what a thrill it was to ride our horses over the new trail right up to the mouth of the Cueva del Agua!

But now came the real problem—removing the idol—and the more we studied the situation, the worse it looked. Even if we could get the idol loose in one piece, we certainly could not carry it. Yet we did not dare break it up by force for fear of shattering it.

I do not remember who suggested a crosscut, two-man wood saw. I was doubtful, but we bought an old one from the Mosqueras and tried it on another stalagmite. It worked—slowly, to be sure, but efficiently; and I learned for the first time that a wood saw will cut limestone. It took Gaspar and Cecilio a little over a week to saw the idol off its base and into pieces small enough to carry out of the cave, out of the hole, and up the cliff to La Patana, where they were loaded on our own pack mules. Within a few hours they were at the lighthouse at Cape Maisí.

With no boards available, there was only one answer to the problem of boxes. We sawed boards out from a cedar log with an old-fashioned pit saw; and we had the nails. For packing, we used Spanish moss from trees in another locality.

In due time, the schooner that brought supplies for the lighthouse arrived and anchored offshore. Don Ramón, lighthousekeeper, sent a messenger, and I rode down post-haste to make the arrangements. The boxes were carried out in a rowboat to the schooner, which delivered them to the wharf at Baracoa, to await the Norwegian banana ship that would transport them to New York.

Once safely received, the pieces were reassembled; and the Idol of Cape Maisí, at long last, was put on exhibition at the Museum of the American Indian, Heye Foundation, where it may still be seen.

No Eyes in the Darkness

A journalistic career that started in Paris and continued in New York City brought William Bridges into association with the New York Zoo, Raymond Ditmars, and caves.

Dispatches to the New York Sun *describing a tropical expedition formed the basis for* Snake Hunter's Holiday, *co-authored by Ditmars and Bridges. Three of the chapters tell of visits to Trinidad caves in search of Vampire Bats and guacharos. Bridges has written nine other books.*

Since 1935 William Bridges has been Curator of Publications for the New York Zoological Society and has accompanied collecting parties to many unusual and faraway places. The first was to central Mexico in 1940, and is described in this chapter. This was the first time scientists had been able to secure a series of blind cave fish which included the eyeless form, the imperfectly eyed form, and the eyed, pigmented form. The fact that this particular blind cave fish came from a warm-water cave made it easy to keep and breed. Much scientific data were garnered from this expedition, and behind the laboratory doors these little fish will continue to help the scientists probe into the mysteries of why, where, what, and how such adaptations for their life in complete darkness evolved.

WILLIAM BRIDGES

I doubt if any of us who took part in the New York Aquarium's expedition to La Cueva Chica back in 1940 had the slightest inkling that it would be said of the blindfish we found there, "no other cave animal has been so successfully investigated," and yet that fifteen years later, one of us, at least, would still be puzzling over them, curious and unsatisfied. Indeed, it would not surprise me if the investigation were still going on fifteen years from now. It looks as if a whole generation of geneticists and endocrinologists and behaviorists will have made a career out of studying *Anoptichthys jordani* and its relatives from a few other caverns and sinkholes in San Luis Potosí.

In the beginning, it all seemed so engagingly simple. There was a cave in central Mexico, and white, blind fish lived in it. We would run down there and explore the cave, find out where the fish came from, bring home a few for study in the laboratory. It might take a little time to work out certain aspects of their biology—that was all.

I am not myself an ichthyologist, and it is easy for me to remember the surprise I felt at the seriousness permeating the meeting that sent us off in quest of *Anoptichthys*. Dr. Charles M. Breder, Jr., at that time director of the New York Aquarium (it was still down at the Battery in those days), called together four or five people who, for one reason or another, might be interested in investigating the biology of those little blindfish in their native habitat. Dr. Edward Bellamy Gresser was one; he was professor of ophthalmology at New York University, and the eyes of fishes fascinated him—even, or especially, fishes without eyes. Ralph Friedmann was there, not for the sake of the fishes, but because there were supposed to be some interesting archaeological remains in that part of Mexico; he liked the idea of teaming up with a scientific party that would provide companionship and yet not get in the way of his digging. Sam Dunton, the New York Zoological Society's staff photographer, sat in because pictures would be important, and I slipped into a back seat to eavesdrop on any potential article for the Zoological Society's magazine. If we actually went to Mexico, we would be joined there by a man from the Mexican Department of Fisheries and by Marshall Bishop, an experienced collector.

Dr. Breder, when he plans an expedition, does not leave any more to chance than is absolutely necessary. I was new to the expedition business in those days, and I thought he went to excess in the mimeographed, five or six pages of items he proposed to lug all the way from New York to north-central Mexico. Alcohol for pickling certain specimens, a boat (collapsible) for getting across pools in the cave, canteens for drinking water, developing solution for Sam's still photographs, electric torches in great plenty—the list went right through the alphabet and overlooked nothing, not even needle-and-thread and cascara tablets.

Skeptical as I was then, I must admit that nearly everything on that list came in handy at one time or another. North-central Mexico and mud-hut native villages, even such comparative metropolises as Valles, with its population of 3,000, where we made our base camp, do not have general stores where one can buy everything one forgot in New York.

Nowadays most tropical fish fanciers have seen, or heard of, the blind Mexican cave fish. It was just beginning to be known fifteen years ago, although Hubbs and Innes had described it scientifically as long ago as 1936. It was a characin—the first blind characin—and the first blindfish of any kind to be named from Middle America. Furthermore, the cave where it lived was a warm-water cave; all other blindfish came from caverns where the pools and running water were cool to cold—55 degrees or less.

The New York Aquarium had acquired a few pairs of fish soon after they were described, and watching these little silvery slivers sliding through their tank, adroitly skirting obstructions and darting unerringly to food dropped in the water, made Breder and Gresser want to know more about them, a great deal more: how they got into the cave in the first place, how they managed to stay there, what they found to eat, what their relationship was to an obviously closely allied characin named *Astyanax fasciatus mexicanus* that lived in the Río Tampaon, a few miles from Cueva Chica. Well, the questions were endless.

Dr. Breder had a report from a Mexican source that should have been reliable, but was not, that access to Cueva Chica was extremely difficult, requiring days of slogging through thick underbrush. He had a better report from one of our own men, Dr. Myron Gordon, a geneticist on the staff of the Zoological Society. Gordon had been collecting in the Río Tampaon the year before and had dropped in on the blindfish cave just to see what he could see. His electric torch was failing, and he couldn't see much, but he did find blindfish in one of two pools only a few hundred feet from the entrance of the cavern. He *thought* the roof sloped right down to the water's edge on the far side of the pool—which would block our progress very early in the game—but he wasn't sure. Anyway, he didn't have to slog through the underbrush. There was a nice, straight, half-mile road heading from the main highway up to a lime-burner's kiln at the very mouth of Cueva Chica.

Dr. Breder was especially curious about what would happen if the blindfish and their eyed relatives of the river were penned up together in the cave and left for, say, a year or two. They would interbreed, undoubtedly, but exactly what would the offspring be like, and how would they behave in darkness and in light? He proposed that, if conditions were favorable, we build pens inside the cave, stock them with eyed and eyeless fish, and come back in a season or so to see what was happening.

All that January afternoon, we listened to Dr. Breder's careful exposition; we talked it over, and the scientists in the group finally decided it would be worth the gamble of time and money to go down to Cueva Chica.

Always, before an expedition, you worry a good deal about things that automatically straighten themselves out once you're in the field. On the way to Mexico I worried about finding the cave, since nobody had ever said *exactly* where it was, or what it was called locally. Furthermore, none of us spoke more than a few awkward words of Spanish. . . .

In Valles, where the map indicated we had better stay, since it was a village of some size, the manager of the excellent hotel was, by a fortunate chance, an American. He knew of half a dozen caves in the neighborhood, any one of which might be the one we wanted, but more to the point, he knew a bilingual guide in Pujal who had explored one particular cave a few years ago with a man from the Mexican Department of Fisheries. Within a couple of hours we were in touch with Sr. Ramón Aguilar, the guide, and our automobile, rented in San Antonio, was headed back up the road from Pujal toward La Cueva Chica—"The Little Cave." This, Aguilar said, was the local name for our cave of the blindfish.

It was as simple and easy as that.

Dr. Gordon had brought back a tiny snapshot of the entrance to the cave where he found the blindfish, and there was no doubt that Aguilar had led us to the right one. As Gordon had said, the entrance was about half a mile off the main road, through palmetto scrub and there was a lime-burner's kiln on the right. The path dipped into the dry bed of a stream, and there before us was the broad, low, rock-roofed mouth of Cueva Chica.

Oddly enough, Dr. Gordon had failed to mention that a dry stream-bed ran into the mouth of the cave. It was significant for our

The entrance of La Cueva Chica. The Indian boy is from the village three-quarters of a mile away, making his daily trip to get drinking and cooking water. *Samuel C. Dunton. New York Zoological Society*

investigation; it meant that at least during the rainy season, a torrent would pour into the cave and carry plenty of food to any fishes trapped in the pools.

As caves go, Cueva Chica was not very impressive or interesting in the beginning. For the first few feet, we could walk upright, then the roof lowered abruptly and skimmed a wilderness of boulders, so that it was necessary to crawl and wriggle. Fifty feet further on, the roof lifted, and we never again had to crawl or even bend low.

Crouching or walking upright, Aguilar led the way, and the rest of us trailed, finding little of sufficient interest to hold us back. There were a few tiny incipient stalactites and stalagmites, a good number of shiny-eyed pholcid spiders, nothing else. In a few minutes, our Mexican guide motioned toward a flaring shelf of gray stone and remarked calmly that we would find plenty of blindfish in the pool under the shelf.

In later years, I have sometimes thought of Aguilar in that moment and have felt sorry that he couldn't experience the thrill we felt as we dropped down in the black dirt and peered over the edge of the shelf into . . . black water and white fish. *Anoptichthys*, in their home. The exultation and intense satisfaction of that moment are not to be forgotten.

Breder and Gresser stared endlessly. They played the beams of their electric torches on and around the little white bodies, and the fish never deviated from their random wanderings. They predicted what would happen if they tossed pellets of mud into the pool— the fish would swim toward the disturbance—and it worked out just as they said it would. The eyeless fish exhibited absolutely no tropism as far as light was concerned, but they reacted unerringly to vibrations in the water. Food that fell into their pool, from roosting or flying bats, for instance, would certainly be found immediately.

It was Aguilar who tired first of watching the ichthyologists play games with the blindfish. He tempted us onward with the statement that there was another pool, and a big one, a very few feet away. It was, indeed, just around the corner of a rock column, and the corridor that led to it was worn so smooth that there was little doubt it was the big pool Dr. Gordon had seen.

There were white fish swimming in that pool, too; not so many, but a fair number. The water, Aguilar said, was fresh and cool and good to drink, so that it served as the dry-season well for a village of Indians three-quarters of a mile back in the bush.

We had the advantage of Gordon in that our flashlights were fresh and their concentrated beams strong and penetrating. In their gleam we could see that he had been mistaken in thinking that the cave ended on the far shore; the walls did close in, but there was a steep-sided corridor in the dim distance, beyond what seemed to be a widening of the pool.

Ralph Friedmann went exploring on his own, poking into crevices and corners in the hope of finding some evidence that the cave had once been used by ancient Indians. He found an upward-turning tunnel, scrambled up it, and came out along the shore much nearer the corridor that we had seen but faintly. Breder joined him, and his shout brought us all. Not only were there white fish in the pool, but crayfish. He had seen one shooting backwards (their normal method of propulsion) from the margin of the pool. It might be a blind crayfish—something new among the cave fauna.

Marshall Bishop waded in, grabbed, and stumbled—perhaps by accident. He was under strict injunctions not to wander out of sight by himself until we had explored the cave thoroughly, but now that he was wet all over, and the water was so pleasantly warm, he kept on going . . . toward the corridor opposite. We saw him haul himself out dripping, and his light faded as he turned a corner. Breder shouted orders for him to come back, and reluctantly, in his own good time, Bishop came.

He hadn't been able to see the end of the corridor, he reported. It was slippery, slimy, and smelly; the water was merely ankle deep, obviously an overflow from the big pool; and in the distance he could hear a sound of running water, almost like a waterfall.

That sounded promising, but it would have to wait. We had not as yet unpacked our gear, and Sr. Coronado of the Mexican Department of Fisheries would be with us the next morning; that would be time enough to set up the collapsible boat and make a real exploration. For our first day, it was enough to know that the blindfish we had come 2,500 miles to find were really there.

By early afternoon of the following day, the "'Pool II Ferry"—in other words, the collapsible boat, set up, dragged into the cave, and launched on the big pool—had made repeated trips across the dark waters, and the necessary collecting and photographic equipment had been landed in the mouth of the corridor where Bishop had made his landfall the day before. The corridor was just as Bishop described it—narrow, slick, slimy, smelling of dampness and decay and bats—but it was easily negotiable. In the second small puddle in the corridor floor, Breder and Bishop simultaneously spotted crayfish. This time the crayfish had no chance of escaping.

A quick examination under the lights showed that they were not remarkable, however; neither blind nor without pigment. They were somewhat lighter in color than the normal crayfish of the outside waters, but the species was entirely familiar.

Bishop could not understand why we didn't hear the rushing of waters that had been so plain the day before. We stood stock-still and listened. A few faint gurglings came to our ears, but that was merely the shallow water slipping between the tumbled rocks around our feet. It was a mystery.

I had undertaken the task of collecting invertebrates in the cave, and there were plenty of spiders and whip scorpions on the walls. Impatient of the little ones, always seeking giant specimens, I unconsciously gained ground on my companions and was soon a hundred feet ahead of them.

Suddenly the corridor turned a corner, and the cave flared and changed character.

Up to this point there had been no fantastic rock formations, nothing but walls and ledges scoured smooth by the rainy-season torrents. Now I had penetrated a chamber that soared sixty feet to a vaulted roof, and the floor became a honeycomb of rock cups, each containing a pool of perfectly clear water. The cups were of all

The third pool seen from the calcium cups during a National Speleological Society expedition to La Cueva Chica. Ernest Ackerly prepares to net blindfish while R. D. Hay watches for bats. *Charles E. Mohr*

sizes, from a few inches to ten feet in diameter; the deepest was hardly more than three feet.

By this time Bishop was close behind me, but the first cup stopped him, for it contained a crayfish. While he collected, I jumped from the lip of one cup to the next and made my way down to the narrowing end of the chamber where it funneled through a low arch.

Casually, I shot my light down the arch and picked out a steeply descending ladder of calcium cups. At their base, fifteen feet below, a pool of dark water was spread like a polished table top. Just under the surface, half a dozen white fish were swimming, but my light caught the glint of something else—tiny, sparkling eyes. A kind of pinkish fire.

Eyes? In a cave of blindfish? I yelled for Bishop; Bishop yelled back up the corridor that Bridges had found something. The restricting walls bounced our voices back and forth, and the answering shouts from Breder and the rest of the party added to the confusion. Bishop caught the excitement in my voice and came leaping; he stumbled against a nest of ants in a log abandoned by the current in the last rainy season, and the ants that spilled onto his bare back and legs turned his shouts into yelps.

It all happened, I suppose, in the space of ten or fifteen seconds. And then I heard a sound that scared the living daylights out of me. It was a roar, a low, dull, pulsating roar that came welling up through the archway where I was standing.

I knew instantly what had happened. The reverberation of our frantic shouting had jarred loose a great plug of rock somewhere below, and the Río Tampaon was rushing into the cave. In a matter of seconds, the cold waters of the river would back up through the arch to flood the calcium cup chamber, and we would all be drowned.

A bat flipped past my head. Another, another, ten more, a hundred more. . . . Just plain, ordinary bats. The roar was the sound of their wings, magnified by the confined space.

I have seen a lot of bats in my time, but nowhere as many as came boiling out of that cavern, through the archway past my head, swirling in the high-roofed calcium cup chamber. They came and they went, sometimes (but rarely) colliding with each other and once or twice smacking into us. There was too much confusion even for a bat's built-in radar, it seemed.

Eventually things calmed down, and with the whole party assembled at the archway we helped each other down the stair-stepping cups to the pool where I had seen eyes. Bishop and Coronado dragged the seine carefully and lifted a wriggling, flopping pocket of fish into the light of Gresser's flashlight. He and Breder stirred them impassively with their fingers, and it says much for Dr. Gresser's self-control that his pipe never went out while he poked, turned, and examined.

For two of the tiny fishes were completely white and eyeless, identical with the ones we had found higher in the cave. Five were white *but* showed dark specks where the eyes should have been. Four were merely pale versions of the colored river fish, still showing traces of their normal markings—and they had big, perfect eyes.

The moment that seine came up and revealed a graduated series of fish, from fully blind to fully eyed, was really the climax of the Aquarium's La Cueva Chica expedition, as far as science was con-

Here are specimens of the fish brought back by the expedition. *Top to bottom*: the blind cave fish, *Anoptichthys jordani;* next, an intermediate form with imperfect eyes; then, a fully eyed cave fish; at the bottom, *Astyanax fasciatus mexicanus* from the open waters of the river—the fish from which the cave-dwelling forms were derived. *Samuel C. Dunton. New York Zoological Society*

cerned. Nobody had ever before discovered a graduated series like that, and to the ichthyologists in the party, the questions of behavior, inheritance, hybridization, and so on, were endless—something that would have to be worked out in the laboratory and not in the field. We had come down asking a few simple questions, and we were going home with a lot of complex ones. The food problem was easy: bats, flies, spiders, water running into the cave from the outside. Where the fish came from was easy, too: undoubtedly they came in through the lower end of the cave, where it emptied into the Río Tampaon. Why they came in, and how they adjusted to survival without light, were the questions that even now have been only partially answered.

A week or two later, after further exploration of some caves and sinkholes in the neighborhood (but not the ones that have since been found to contain other, related blindfish), and after making stills and a motion picture in the cave, the expedition was ready to return to New York. None of us had realized, back there in Dr. Breder's office in New York, what the trip would mean to us.

At the personal and physical level a price was paid by most of us. Dr. Gresser and Marshall Bishop were suddenly stricken with high fever; Sam Dunton came down with it; Ramón Aguilar felt it. Ralph Friedmann had made an early departure for New York and was attacked after he got home. Only Dr. Breder, Sr. Coronado, and I—who had shared food, beds, and cave work with the stricken ones—escaped without a touch of fever.

What it was, I think nobody really knows, even now. It cost Gresser and Bishop weeks in the hospital on their return, and the best medical men in New York pulled them through. That is all I know.

Somehow, even though the transportation of very sick men was our first consideration, we managed to bring back to New York an adequate collection of live specimens of all forms of the fish, from eyeless to eyed. That was in 1940; successive generations are still appearing regularly in Dr. Breder's tanks at the American Museum of Natural History, where he is now chairman of the Department of Fishes. One of his first experiments was to build a laboratory "cave," an aquarium twenty inches wide, forty inches long and four inches deep, one half of it covered to create darkness and the other half exposed to the light. Some interesting points emerged. Fully blind fish showed a slight preference for the dark end of the tank, as did their fifth-generation descendants that had been reared in light. Partially eyed fish—those which could form some sort of retinal image—had a strong schooling instinct, which was lacking in the completely blind. Individuals with vision tended

Three blind cave fish, *Anoptichthys jordani*, from La Cueva Chica, Mexico, brought back by the New York Zoological Society and exhibited at the Zoo. This is the species that has been successfully bred in captivity and which is popular among collectors of tropical fish. *Samuel C. Dunton. New York Zoological Society*

to rest quietly in a school, while the blind fish kept up an apparently aimless wandering. In light, the eyed individuals often attempted to school with the blind fish—and these attempts frequently ended with the eyed killing the blind.

Drs. Breder and Gresser commented, in a paper about those early experiments: "The differential behavior existing between the blind and the seeing is apparently an additional positive factor in the establishment of this cave form."

Since then, additional collecting and exploration in San Luis Potosí and the adjoining state of Tamaulipas has revealed four other caves containing blindfish: Cueva de Los Sabinos, Sotano de la Tinaja, Sotano de la Arroya, and Cueva del Pachón. At least two other blindfish from them have been described: *Anoptichthys antrobius* and *Anoptichthys hubbsi*. With more material to work with, the problems have increased, rather than diminished.

Since the cave forms are obviously derived from the common little characin of the open waters. *Astyanax f. mexicanus,* much of the investigating has started from it, in order to see how it could or did adapt itself to cave conditions. But it seems that *Astyanax* develops some very serious hormonal upsets when it is kept in the dark, and the pituitary, thyroid, and gonads are affected. Remove the eyes from a river fish but keep it in the light, and it does not develop this imbalance. The naturally blind cave fish do not fall prey to these difficulties when they are kept in the light. It seems certain that a definite endocrinological adjustment had to be made before the ancestral *Astyanax* could exist in a cave. In short, there is nothing "degenerative" about the evolution of blind characins; rather, there had to be a positive adaptation to absence of light.

Experimentally, it has proved very difficult to bring about this positive adaptation—the fault, perhaps, of trying to do in fifteen years what nature may have taken ages to accomplish—and various deformities and even cancers have been found in the fish undergoing experiment.

As Priscilla Rasquin and Libby Rosenbloom put it in a recent paper (*Bulletin of the American Museum of Natural History,* Volume 104): "The fact that cave fish derivatives of the species studied do exist proves that at least two fish succeeded in making a transition to life in total darkness. It is clear that all individuals cannot survive the transition."

It had all seemed so engagingly simple when we knew so much less about it.

Ozark Cave Life

The underground features of the Ozarks have always held an attraction for explorers with scientific training. The deep beds of limestone in which hundreds of caves occur contain minerals of considerable interest and worth. But it has been the discovery of new strange denizens of underground rivers and lakes in the Ozarks that led to a procession of excited though sporadic pilgrimages—herpetologists hunting salamanders, ichthyologists studying blindfish, comparative anatomists seeking for the causes of blindness, and many other zoologists, biologists, and naturalists.

This is the story of the findings of just a few of the speleologists and of the author's experiences on two of five trips he has made to this rich but still only partly known portion of our American underworld.

Charles E. Mohr

U. S. Highway 66 strikes southwestward from St. Louis and rises gradually in graceful curves into the heart of the Ozark Mountains. For three hundred miles it winds among gently rounded hills and prominent limestone cliffs until the landscape levels out again not far from Tulsa, Oklahoma.

The highway passes within an hour's drive of hundreds of caves, nearly twenty of which have been lighted and otherwise prepared for public inspection. This region of cavernous ridges and castellated cliffs spreads across southern Missouri, northern Arkansas, and a small portion of northeastern Oklahoma.

Four hundred years ago, Hernando De Soto crossed the southern Ozarks after discovering the Mississippi River. He failed to find the mineral treasures he sought there, but French prospectors two centuries later, and the Spaniards who followed them after the ceding of the territory to Spain in 1762, were more fortunate.

They came seeking treasure in a region where "the lands were reputed to equal in fertility the banks of the Nile, and the mountains to vie with the wealth of Peru." In the deep layers of limestone they found riches, not in the form of gold and silver, but in vast deposits of lead and other lesser known minerals.

Certain of the caves of the region must have been known to the earliest settlers, because when Henry Schoolcraft explored the eastern Ozarks in 1818 to study the "lead-bearing limestone" for the United States government, he found saltpeter workings. The young geologist—who became famous for discovering the source of the Mississippi, and for his five-volume work on legends and traditions of the Indians—often camped in caves. Though he was later to become a champion of the Indians, Schoolcraft feared the hostile tribes as he forced his way through the narrow valleys and uncharted forests of the Ozarks.

The caves he used for shelter stirred his fancy and aroused his scientific curiosity. One cave he found at night, during a storm. After building a fire in its welcome shelter and cooking supper, he lit a torch and set out to "explore the recesses of the cave, lest it should be occupied by some carnivorous beasts, who might fancy a sleeping traveller for a night's meal."

Schoolcraft wrote that he was alert to the possibility of finding "animal bones in our western caves, as those of Europe had recently excited attention; but never found any, in a single instance, except the species of existing weasels and other very small quadrupeds."

After the Civil War, Missouri became the jumping-off point for settlers heading west, but the mineral resources of the southwestern Ozarks also attracted prospectors and speculators. It was in this area that reports of a very large cave began to be circulated.

A party of miners searching for lead ore had succeeded in entering a huge cave through a hole in its roof and finally had been stopped at the edge of a bottomless abyss. Known as Marble Cave, it was situated about forty miles south of Springfield. About 1884 the Marble Cave Mining and Manufacturing Company was organized to handle the "marketing of the millions of tons of guano, marble, onyx, emery, granite, and other valuable material hidden inside Roark Mountain." One of the officers, Truman S. Powell, pioneer newspaper editor, later leased the cave. Early in the twentieth century he became famous as the prototype of the Old Shepherd in Harold Bell Wright's best-seller, *The Shepherd of the Hills*.

Nation-wide interest in Marble Cave was stimulated by an article that appeared in the *Scientific American* in 1885. Fourteen rooms or passages were described, two of which should have alerted scientists: "A fifth passage, the most tortuous of all, circular in shape, about 190 feet long, and at some points only two and a half feet in height . . . is filled with mummified animals. There are bears, panthers, otters, raccoons, opossums, wolves, foxes, lynxes, etc., and one specimen of what seems to be an antidiluvian animal of the genus *Pterodactylus*. Also, smaller animals, seemingly some kind of monkey. These mummies are in a state of repose, as if the animals had come here to die. Hence the room is called the Cemetery. Hair on the dried up skins is well preserved. . . .

"Waterfall room is crescent-shaped, and about 100 feet high. Down it falls with great force, a large body of water coming from rooms above. . . . In front of the fall are jug-shaped basins . . . containing white, soft, blind, toothless lizards, from two to six inches in length."

The description of the fossils and of a "white, blind lizard" should have sent scientists scurrying for the Ozarks, but either they were preoccupied with Mammoth Cave's fabulously abundant and varied animal life, or they considered the Ozarks still too unexplored and remote to justify a trip to investigate this astonishing rumor. It was not until six years later that one of the "lizards"—actually a salamander—was finally collected (from another cave in the area), pickled in alcohol, and sent to the Smithsonian Institution in Washington. There the chief zoologist, Leonhard Stejneger, examined it and published a description of it. The only other creature of this type, the Olm of southern European caves, was the most famous of blind cave animals.

"The discovery of a blind cave salamander on our continent," he wrote, "is one of the most important and interesting historical events of recent years. . . . The interest is considerably heightened when we have to do with the first and only blind form among the true salamanders."

The blind amphibian had been sent from Rock House Cave, near Sarcoxie, Missouri, but efforts to capture additional specimens there were unsuccessful. Stejneger's description, while mentioning the small, slightly rounded eyes concealed under the continuous skin of the head, was incomplete. A detailed anatomical study was not attempted, he wrote, "as I have not felt justified in mutilating the type specimen. . . ." The color of the animal was described as "yellowish." He named it *Typhlotriton spelaeus*, utilizing the Greek word for cave, *spelaeon*.

No scientist had ever seen a living blind salamander in this country, so in 1893, famed naturalist and comparative anatomist Edward Drinker Cope journeyed to the Ozarks, visited with Truman Powell at Marble Cave, and spent two days exploring.

"I obtained several specimens of this interesting species . . . in spirits it has a pale yellow color, as described by Dr. Stejneger, but in life it is white. It occurs in a stream that flows at least 300 feet below the surface."

While exploring the surface streams of the regions, Cope found a long-tailed salamander unknown to science, and described it as *Splerpes melanopleurus*. When anatomist Carl Eigenmann came to the Ozarks in 1897 to study blindness in cave animals, he found Cope's salamander in a cave. Supposing it to be an unknown variety, he also published a scientific description and named the salamander in honor of Dr. Stejneger. The duplication of names was not discovered until many years later. Now only the original description and name, given by Cope, are recognized. The salamander shows no adaptation to cave existence.

Preparations for the St. Louis World Exposition scheduled for 1904 to mark the centenary of the Louisiana Purchase, were getting under way in the early nineties. Natural features of the area were being assessed, and Marble Cave was included among the places visited. Geologist E. O. Hovey, who headed the survey party, was so impressed with the cave that he prepared and illustrated an article, which was published in 1893 by the *Scientific American*.

Hovey told of seeing long, notched poles near the entrance, which were believed to have been used by Spaniards who "occupied the land in this region before the English settlers took possession of it. . . ."

"Some light comes through the great rift in the roof, which is the *bottom* of the sink hole, and as soon as our eyes became adjusted to the semi-darkness we could see something of the really grand dimensions of the immense dome in which we stood; but when the room was illuminated by red fire, its full grandeur was revealed. The room is about 150 feet wide by 200 feet long, and the roof rises in a magnificent arch to a height of 165 feet from the floor."

The Great White Throne, fifty feet high and fifty feet wide, had special appeal for Hovey. It showed, he said, "all the beautiful forms which one might imagine to be caused by the freezing of a fountain."

A few feet away was a fine spring. From this spot, a narrow passage continued to the Animal Room. Hovey repeated the story that this room contained "the mummified remains of hundreds, even thousands of animals, mainly, if not entirely, of carnivorous species." "Admittance to this room is positively forbidden by the owner of the cave," he reported, "but assistants from the Smithsonian Institution at Washington have had access to the material from it and are now at work on their identifications. A specimen from this room which was shown to me consisted of the skull and jaw-bones of a cat-like animal to which portions of dried skin and fur still clung."

Oddly enough there seems to be no existing record of the bones sent to Washington, or any present knowledge of where the Animal Room is located.

Eigenmann, on his trip to the Ozarks a few years later, was unable to find the chief object of his search, an adult blind salamander. He succeeded in collecting a few of the young ones. The next year, knowing that the rare adults had been found in Marble Cave, he returned and became the second scientist to see living adults of the unique white amphibian.

The specimens which he brought back to Indiana University, long a center of bio-speleological research, provided the basis for some of the most important data published in his classic bulletin *Cave Vertebrates of America, a Study in Degenerative Evolution.*

Eigenmann had discovered young or larval *Typhlotriton* living in springs at the mouths of caves where they were exposed to light. When he studied the minute structure of their eyes, he made the surprising discovery that the larvae had nearly normal eyes. But the eyes of adults showed evidence of almost complete degeneration. The rods and cones of the retina had disappeared, and the eyelids had overlapped and become fused together.

When did this degeneration take place? he wondered. Examining salamanders in different stages of development, he found that the eye structure began to break down while the larva was transforming to adulthood, a time when many other important physiological

This close-up of an adult Ozark Blind Salamander clearly shows the fused eyelids. The adult is able to leave the water and travel about in the humid atmosphere of a moist cave. *Charles E. Mohr*

changes were taking place. It was then that the larva's external gills disappeared, and henceforth respiration could take place only through the salamander's moist skin. A dense network of capillaries located close to the surface of the skin permitted the exchange of carbon dioxide and oxygen as long as the skin remained moist. Unlike the aquatic larva, the adult was able to leave the water and travel about in the humid atmosphere of a moist cave. If exposed to heat or dryness, however, the creature would perish quickly.

Eigenmann's reports evidently satisfied for a time the curiosity about these strange salamanders. Thirty years later, with the rise of experimental biology and increased interest in the role of the endocrine glands in growth, *Typhlotriton* again became a challenge. What part did the cave environment play in bringing on its blindness?

In 1926, the American Museum of Natural History's herpetologist and experimental biologist, G. Kingsley Noble, explored a

The larva or young of the Ozark Blind Salamander, *Typhlotriton spelaeus,* has functional eyes and breathes through external gills, visible in salamander at left. The adult at right has lost both gills and eyesight. *Charles E. Mohr*

number of Ozark caves, including Marvel Cave, as Marble Cave came to be known.

Leaving the well-traveled tourist trail behind, he began to explore. "We came upon our first *Typhlotriton* shortly after beginning the crawl," he relates. "It was walking rapidly over the loose stones in an effort to escape. In the light of the gasoline lantern its pale tones gave it a most unreal appearance. The beast looked more like some creature fashioned out of dough than a living organism. Its dark eyeballs shone through the translucent lids and gave it a peculiar expression. But once the creature was in the collecting tin it appeared more animate. Close scrutiny revealed that its skin was not devoid of pigment. The net of capillaries underlying the epidermis was covered with melanophores, but these pigment cells were fully contracted. The creature reminded me of the salamanders whose pituitaries I had removed in the laboratory, and I could not help but wonder if the pituitary body was functioning normally in this creature."

Continuing his search for salamanders outside the cave, Noble came upon many larval *Typhlotriton* in springs. Not only were their eyes normal, but they were dark-skinned, heavily pigmented. He also found larvae of a long-tailed salamander, but rarely in the same springs or streams. Where the water temperature ranged below 65 degrees Fahrenheit, *Typhlotriton* occurred. Where the water flowed for a distance under the surface rock and had warmed to 70 degrees, the long-tailed larvae were predominant.

In this way, the two species avoid competition, Noble concluded. "The species which inhabits warm water metamorphoses without losing either its eyesight or pigmentation, but the one that lives in cold waters undergoes a remarkable change."

The scientist noted that laboratory experiments were throwing "much light on the nature and mechanism of metamorphosis. The role of the thyroid hormone in producing metamorphic changes is well known, but the importance of external factors in altering the end result is not so clear. Thus, it is possible that *Typhlotriton* raised in waters of different constitution or temperature might show a greater or lesser degree of these metamorphic changes. The immediate problem of the importance of external vs. internal factors in producing the ghostly creature found in the caves is removed from the domain of field observation to that of laboratory work. The adult *Typhlotriton* have never been found outside of the caves, and yet each generation of larvae spends two or more years outside of the caves as well-formed and pigmented salamanders. The destructive changes of metamorphosis necessitate a retreat to adjacent caves, where a combination of external and internal factors results in the animal adapted to cave life. Generations of cave life have not affected the form of the larva and, without laboratory analysis, who can say that it has permanently affected any feature of the animal's heredity, even its metamorphosis?"

Could he demonstrate that the loss of sight was the result of life in cave surroundings? Noble wondered. He took his collection of live salamanders back to the museum laboratory in New York. He divided the largest larvae into two groups. One group, he kept in cool, moist, cave-like conditions, in total darkness. The other group, he exposed to the light.

After six months, the salamanders had transformed. Those reared in the dark showed typical degeneration of the eye, while those reared in the light retained and further developed the functional eye possessed by all larvae; they also developed more pigmentation. The many generations of cave life had not affected the animal's heredity.

I had read Noble's fascinating account of these "Creatures of Perpetual Night" in *Natural History* magazine while I was still in college and had talked with him about further experiments he planned. The project was interrupted, however, and never resumed.

For years I dreamed of visiting the Ozarks and seeing the blind salamanders. My chance finally came on a trip to St. Louis. It was Christmastime and I had a couple of extra days. I sped to Marvel Cave through powdery snow, to be greeted warmly by the Lynch sisters, who owned the place.

They led me down a long wooden stairway to a platform overlooking a vast, black space—a hole in the dome of the unlit ball-

A forty-foot wooden tower is built on top of an underground mountain in the entrance room of Marvel Cave, Missouri. *Charles E. Mohr*

room—then down a tall, wooden tower to the top of an underground mountain.

Halfway down, I spotted a formation that looked remarkably like an owl. When I put my light on it, the "stalagmite" came to life and flew out of the cave! It was a Screech Owl. We followed the descending trail till we reached a dripstone canopy; circling behind it, my guides pointed to a small spring.

"Sometimes we see salamanders here, but only young ones." There were several half-inch-long, white isopods in the pool and a larger, pale creature—a blind salamander! A larva! It had eyes, and there were feather-like gills protruding from its neck.

Did adult salamanders still live in more remote parts of the cave, I asked? Yes, they were still to be found farther back in the cave, beyond Blondie's Throne. "Young Charlie Davidson crawled back there while he worked for us; if he's at home you might get him to be your guide."

Following their directions through Harold Bell Wright's *Shepherd of the Hills* country, down the road past "Preachin' Bill's" cabin, I found Charlie chopping wood. He was willing, so early the next morning we set out in search of *Typhlotriton*.

Our trip really began at the waterfall. A fifteen-foot, slippery clay bank stopped us until we had dug toeholds. At the top, we found ourselves in a series of pools of varying depths. This was the Mystic River I had read about. The first plunge into 55-degree water left us breathless, but our exertions soon warmed us. We alternately waded and crawled beneath the low, level ceiling, scanning the passage for specimens, but none could be found.

At last Blondie's Throne was above us. Charlie Davidson told me that it was a small, beautifully decorated room named for a fair-haired boy who found it long ago. I was too eager to find a salamander to spare the time necessary to climb to the celebrated chamber. Ahead, a broad pool marked the continuation of Mystic River. We waded knee-deep along the shallow edge, looking hopefully for salamanders, but still without luck.

Pushing on, we searched more anxiously, down on our knees again as the roof abruptly lowered. The stream slowly dwindled and finally disappeared. A passage to one side looked promising, but it soon became twisting and narrow and began to rise steeply. Loose rock made crawling a clumsy process, and I was breathing with difficulty. After another squirming climb and sharp turn, the passage opened into a low room. My breathing became even more labored. I was perspiring freely. Was this the famous "Cemetery" or Animal Room? Could it be one of the oxygen-deficient pockets said to occur in a few caves?

I wasn't equipped to make observations on atmospheric conditions, so with the blood pounding in my ears, I backed awkwardly down the crooked, narrow passage. Back by the underground river, I gasped in the cooler air. The impulse to hurry out of the cave was nearly overpowering, but just then, directly ahead of me, I saw a blind salamander poking its head over a small ledge.

"It's looking right at us," I whispered excitedly to Charlie, but he corrected me. "It's blind, isn't it?"

It was completely without color, "like grass that had a plank over it too long," Charlie said. It moved along like a sniffing dog, apparently in search of food. Scarcely daring to take my eyes off it, I quickly unpacked my photographer's kit, set up my camera on a tiny tripod, and in the instantaneous glare of my flash bulb caught the alert pose of this strange, sightless salamander. It showed no reaction to the intense light.

We encountered several more *Typhlotriton* on the way out. Had we overlooked them on the inward trip? Or were they like a colony of long-tailed salamanders I studied years later in Pennsylvania— so full of curiosity that I could depend on them to come to the entrances of crevices and burrows and investigate me?

Where and when do these salamanders lay their eggs? Since neither the freezing temperatures of winter nor midsummer's arid conditions can reach them, they probably could reproduce at any season. Even today no one knows the answers. No one has found the eggs of *Typhlotriton* in nature.

James Kezer made numerous trips to the southern Ozarks to ferret out the life history of *Typhlotriton*. Finally he found several female salamanders which plainly showed large eggs through their semitransparent skin. Surely they must be nearly ready to deposit them beneath some rock along the cavern stream. But Kezer did not wait for that to happen. He took four of them back alive to the laboratory.

There Kezer and his colleague, Robert Barden, decided to try to stimulate egg-laying by the use of hormones found in the pitui-

The Cave Salamander, *Eurycea lucifuga*, has well developed eyes and lives in the twilight zone just inside cave entrances. It frequents caves from Virginia to Oklahoma. *Charles E. Mohr*

tary gland. This they did by removing glands from other species of salamanders and from frogs and implanting them beneath the tongues of three *Typhlotriton*.

After a week during which six salamander pituitaries and one from a frog had been transplanted, one *Typhlotriton* laid four eggs. Eleven days later, following additional transplants, a second batch of eggs was deposited. The scientists watched the salamander as she moved over the moist surface of a flat rock, laying eggs just above the level of the water.

The eggs were cream-colored, barely a tenth of an inch in diameter, but surrounded by transparent layers of gelatine, so their total diameter was about one-third of an inch. Deposited singly, the eggs stuck to the moist rock. The small collection was preserved. Subsequent efforts to find eggs in nature or to carry other embryos through to hatching have been unsuccessful.

Nothing further is known about the early development of *Typhlotriton,* nor have the eggs of the still rarer Texas Blind Cave Salamander, *Typhlomolge*, ever been found. A third species of American blind cave salamander was spewed up from an artesian well in southern Georgia in 1939, but only the single specimen has ever been seen.

A few miles from Marvel Cave, I first saw the Ozark Blind Fish and found an unusual white crayfish. I had asked everywhere about eyeless fishes, hoping to find the species which a cave explorer named Ruth Hoppin had discovered near Sarcoxie in 1889. She had found them in well-like caves, so when I heard of a cave of this type near Springfield, I set out to locate it.

Kenneth Dearolf and I found the cave in a pasture. The farmer had filled its manhole-like entrance with sticks to keep his animals from falling into it. Quickly clearing out the sticks, we examined the opening and saw that it was a narrow chimney, descending to a ledge about twenty feet below. The descent was easy.

Ken lowered the equipment to me on a fifty-foot length of three-eighths-inch rope, then came down himself. A small stream flowed past us, then plunged into a fair-sized room, splashing on a rock pile some twenty feet below. The only place a rope could be fastened was close beside the waterfall.

We were unprepared for a descent; we didn't have a second rope with us to use for a safety. Rather than take the time to go back to the farmhouse for one, I decided to go down without a safety, since the drop was such a short one. It was a foolish decision; safety ropes should always be used.

I doubled the rope and tied knots at convenient intervals. Anchoring one end, I dropped the free end over the edge. It didn't quite reach the floor—but it seemed only about a foot or so short. I started down.

The walls curved away. With nothing to hold the rope taut, it began to swing, and halfway down, the swinging line carried me under the waterfall. In a moment I was drenched. Chilled and blinded, I clung desperately to the rope, but the water poured into my hip boots. The sudden increase in weight was too much, and I lost my grip on the rope. I landed on the rocks directly beneath the waterfall. Choking and gasping, I crawled away from the pounding water into the darkness.

Out of reach of the water, I stopped to survey the damage. My glasses were intact, and my flashlight still worked. I had incurred nothing worse than bruises. Reassured to see me moving about, Dearolf lowered my collecting kit.

I was in the bottom of the well, about twenty feet across, but the water was flowing off to one side, into a small, low room behind the falls. Water covered most of the floor, flowing into a stream passage too small to follow. The water was somewhat muddy, but I could see

This Ozark Cave Crayfish carries its eggs until the embryos are well developed. This cave species, *Cambarus ayersii*, attains a length of three inches. *Charles E. Mohr*

white crayfish crawling about. The largest seemed to be carrying something beneath its abdomen. I succeeded in netting it. Transferring it to a jar, I could see that it was a large female carrying several dozen white eggs.

I was knee-deep in the pool when for an instant a white fish appeared, then sank right out of sight again. For minutes I stood motionless, hoping for another view of it, but the water was now so muddy I couldn't see anything.

Suddenly I became conscious of the roar of the falls. It seemed much louder now—and the ceiling of the small room seemed a little lower. The water was rising! I heard Dearolf shouting but couldn't make out what he was saying.

Unknown to me, it had begun to rain quite heavily. A lot of water was coming into the cave. Dearolf was pointing upward; time to get out, he seemed to be saying. I headed for the rope. The lowest loop came just to my shoulder. Weighted down with water-filled boots, I couldn't pull myself up. I took them off and tied them and the rest of my equipment to the end of the rope. Ken hauled them up, then dropped the line down to me.

Even without the boots, I couldn't make it. I shouted to Ken to go to the farmhouse for a heavier, longer rope but he couldn't hear me above the watery bedlam. At last he seemed to understand my sign language and disappeared.

I ducked into the small room again. It wasn't so cold here, for there was less circulation of air. I turned out my light to save the batteries and swung my arms to keep warm.

The water kept inching upward. Finally, after what seemed an hour, I left the room just as a sturdy rope dropped into view. The light rope, being doubled, was still too short for a safety, so, taking a deep breath, I started to climb the heavy one. Wet as I was, the sudden deluge under the waterfall was again a shock. The force of the water jammed my headlight down over my ears; that was all that kept my glasses on. Halfway up, the pounding began to slow my climb. A few more feet and I stopped. I clung to the rope, expecting to be dashed down at any moment. While swinging back and forth, my foot touched a ledge. Kicking the ledge, I used the rope as a pendulum, swinging in a wider arc until I was able to find footing on the ledge out of the water's reach.

I shouted to Dearolf to let him know what had become of me. I was just beneath the brink of the falls. Only a few feet to go. Taking a vise-like grip on the rope, I swung out into the cascade again. Climbing mightily, I quickly pulled myself up through the water. Dearolf was there to grab my arm and pull me to safety.

We climbed to the surface, to find that the storm was about over. It had been a veritable cloudburst. A trickle we had stepped across in the meadow had become a broad torrent, and dozens of drowned chickens were floating past. The level of the main flood was only inches below the cave entrance.

Three weeks later we returned. This time, we anchored a wooden beam several feet from the falls and hung our rope from it. But we were drenched anyway. The whole floor of the cave was covered with water, though I still could wade. And the fish was there. This time it didn't get away.

It was tiny and completely white, scarcely two inches long. We sent it to ichthyologist Carl L. Hubbs. Had we discovered a new species? Eventually word came from Dr. Hubbs. It was *Troglichthys rosae,* the blindfish Ruth Hoppin had found nearly fifty years earlier. Dr. Eigenmann had named it for his wife, Rosa.

Like Miss Hoppin, we noticed that the fish was sensitive to vibrations in the water, or even tapping on the rocky wall of the pool. Sounds, however, did not affect it. She had written: "I tested their hearing by hallooing, clapping my hands, and striking my tin bucket when they were in easy reach and near the surface. In no case did they change their course or notice the sound."

The ability of the fish to detect even slight disturbances of the water evidently lies in the papillae, each tipped with a sensitive nerve ending, which are arranged in rows over the top and sides of the head and jaws. There are prominent eye-like structures, but examination proves them to be round masses of fatty tissue, in which the minute, fully atrophied eyes are buried.

"Judging from the degree of degeneration of the eye," Eigenmann wrote, this Missouri fish "has lived in caves and done without the use of its eyes longer than any known vertebrate."

Eigenmann's studies put an end to excited speculation that this blindfish might be identical with ones found in Kentucky caves. At that time, scientists could not imagine that an identical cave fish could turn up in two widely separated cave regions like central Kentucky and the western Ozarks, especially since they were divided by the Mississippi River.

But in 1954 Loren Woods and Robert Inger of the Chicago Natural History Museum found that the common blindfish of Kentucky and Tennessee actually does live in Ozark caves. They believe that since the same massive beds of flat-lying cavernous limestone that reach the surface in Kentucky and Missouri pass under the Mississippi River, interconnecting channels covering hundreds of miles may have existed in the past or may still permit the circulation of underground waters and their tiny inhabitants.

Few blindfishes in the United States reach a size of three inches, but their small size is not attributed to a limited diet. Insects, such as cave crickets, flies, and fungus gnats, and bats that fall into the water, provide considerable food, along with aquatic creatures such as young crayfish and tiny, white, eyeless crustaceans like the flattened isopods and thin-bodied amphipods.

Where debris finds its way into caves, or where bat colonies provide organic wastes, life may exist in abundance. The most notable example I have ever encountered was at the western edge of the

Bat tapestry in Marvel Cave, Missouri, is composed of 14,500 hibernating Gray Bats, *Myotis grisescens.* Charles E. Mohr

Ozarks, in northeastern Oklahoma. We were following along a cliff pock-marked with waterfalls, springs, and small caves, and came at last to one known locally as Bat Cave.

Inside the cave, we found our progress blocked by a "guano bog," a deep deposit of bat droppings over which a stream was flowing. Probing it with a pole, we could find no bottom. On the far side of the stream, guano extended fifteen feet above the water. This was the bat roost. Hundreds of bats that had been flying about as we approached now disappeared into distant parts of the cave.

The occupants of the quagmire fascinated us. In other caves, we sometimes found a cave flatworm or two, rarely as many as a dozen. Here were *hundreds* of the white, blind flatworms. Several fair-sized concentrations were along the edge of the bog, where they could be photographed.

As we looked elsewhere in the soupy liquid and beyond in the clear water, we saw them by the thousands. The total colony must have numbered in the tens of thousands. Along with them were hundreds, possibly thousands, of isopods. Some were clustered on a submerged bat, evidently feeding on it.

The flatworms, or planarians, as they are known to biologists, were concentrated on smaller objects, such as dead insects and pieces of disintegrating wood. Either directly or indirectly, however, they must take nourishment from the guano.

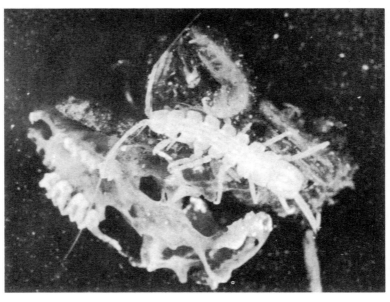

An isopod clinging to a bat skull in a Marvel Cave pool. *Charles E. Mohr*

The planarians ranged in size from minute, thread-like creatures a few millimeters long to specimens measuring nearly half an inch. Often they flowed along on the surface film, or settled onto the guano and moved by alternately widening and lengthening their bodies.

The flatworms are about half an inch long and can swim at the surface *(group at right)* or move along the bottom in Bat Cave, Oklahoma. *Charles E. Mohr*

Although they are related to the internal parasitic liver flukes of higher animals, these cave planarians are free-living and harmless.

The fauna of caves may be roughly grouped into three categories. Permanent cave dwellers are those like the white planarians, and the blind amphipods, isopods, and white crayfish, and, of course, the eyeless fishes and adult blind salamanders. All of these are unknown except from caves. To be sure, they may be brought to the surface from deep wells, but they swim or are carried upward from areas which are cavernous, sometimes hundreds of feet below the surface. The environment in which these creatures spend their entire lives is about the most unvarying that can be imagined—though it is approximated by the black depths of the oceans. Truly cavern-living species exhibit radical structural modifications as a result of cave existence, just as the dwellers of the deeps differ from those who live nearer the surface of the sea.

Some cave inhabitants which spend much or most of their lives underground may be equally at home in other moist, dark situations, such as in cellars or beneath rocks or logs. Except for a few true

The familiar cave cricket is really a voiceless, long-horned grasshopper. It often leaves the cave at night to feed outside. Many species of *Ceuthophilus* have been found in caves throughout the United States. *Charles E. Mohr*

cave-dwelling species, cave crickets live near enough to the surface to venture forth at night in search of more abundant food supplies. This is true also of several species of eyed cave salamanders, of flies and moths, of pack rats, and of the bats which may pass the winter underground, or which may occupy cave "maternity quarters" in early summer.

Still other forms of animal life retreat from the surface to the twilight or entrance zones of caves when opportunity affords or occasion demands. In northern Arkansas, for example, I visited caves during hot, dry weather and sometimes found dozens of the silver-flecked, stout-bodied Slimy Salamanders resting on the walls and ceilings. The Ozark Red-backed Salamander is another species which often may be found in caves. The relatively constant conditions of moisture and temperature in most caves provide welcome refuge.

Some of these salamanders doubtless breed in caves. Eggs of the Slimy Salamander were first found in an Ozark cave, while the only place where the Red-backed Zigzag Salamander is known to lay its eggs is in Mammoth Onyx Cave, Kentucky. The eggs of most cave-dwelling species have never been found.

So much remains to be learned about the life cycles, habits, migrations, and evolution of cave animals that a generation of biologically trained observers could profitably devote their best efforts to the solution of these problems. The Ozarks, with their abundance of caves and wealth of cave life, offer a fertile field for the study of salamanders, eyeless fish, and white crayfish, and of the smaller but more abundant forms of life without which the larger creatures would quickly starve.

The Leather Man

Some years ago, Leroy W. Foote set out to investigate three caves he heard had been inhabited at one time or another by a hermit known as "The Leather Man." Foote originally planned to survey them on three consecutive weekends. Instead, it took him six years to finish the job.

During this time he interviewed over four hundred people, most of them old enough to have remembered the Leather Man. His three caves became thirty-four; sixteen old photographs of the Leather Man were uncovered in family albums; even life-size water color and oil paintings of this queer man were found. By the time Foote was finished, he had accumulated a bulging file containing Leather Man data never before recorded, and he had examined many of the odd contents of the Leather Man's pack.

Leroy W. Foote

Sometime early in the last century, a son, Jules, was born to a young couple in Lyons, France. The boy got his education at the local *lycée,* studying hard in preparation for the economic struggle ahead of him in that strike-ridden, silk-manufacturing city.

Shortly after leaving school, he fell in love with Marguerite Laron, the attractive daughter of a wealthy leather merchant. She returned his affection, but her father opposed the match, objecting to the difference in the social level of the two families, for young Jules Bourglay was the son of a wood carver, several rungs down the ladder from a leather merchant.

Laron's objection brought matters to a standstill, but finally the young man won an interview with him. As a result Jules went to work for the leather firm, with the understanding that if he did well during the course of a year, he could marry Marguerite and become a member of the firm.

Before long he was confidently practicing the intricate business of buying leather, but just when his future looked brightest, a depression hit the city. Almost overnight the price of leather dropped 40 per cent. Poor Jules lacked the experience to cope with the emergency. An over-purchase of Persian leather in this falling market pushed the firm into bankruptcy, and Jules found himself without a job, his hopes of marrying Marguerite gone forever.

The shock and disappointment were so great that he was under a physician's care for the next two years. Then he disappeared and was never again seen in France.

East of the the Hudson River's Tappan Zee and forming the eastern ramparts of the valley are the Taconic Hills. For nearly a hundred years the folks who live throughout these hills have repeated fantastic stories about a wandering hermit; tales that persist to the present time, for even now old-timers will point with convincing pride to some sheltered valley and tell of a cave-dwelling man who frequented the spot.

Within this area between the Hudson and the Connecticut rivers, lying north of Long Island Sound and south of an uncertain line near the Connecticut-Massachusetts boundary, are scores of small caves and rock shelters. Several dozen of them have long figured in the local folklore and are known by such picturesque names as Tories Den, Dug Way Cave, Deep Hollow Cave, Gamaliel's Den, Turkey Mountain Cave, Faun Rock, Totoket Cave, Jericho Rock, and the Huckleberry Hill and Ecubut Hill caves.

Spaced about ten miles apart, these rock shelters form a great circle within the larger area and help to perpetuate a strange, romantic history that once fascinated the folk of southern New England. For here traveled a solitary man whose story is a compound of legend and reality. He was known as the Leather Man, and the two score natural shelters he used were known locally as Leather Man caves.

At regular intervals for thirty years, fires burned at these cave entrances, so regularly, in fact, that local residents knew exactly when to look for them. Smoke rising from a Leather Man cave was always a subject of conversation, but it surprised no one. The absence of fire or smoke at the proper times would have been cause for serious speculation, because each Leather Man cave was an object of local interest and responsibility. Always watched, often visited, seldom molested, and continually a source of wonder, each cave was treated with genuine respect.

Occasionally, even today, fires are seen before certain caves, but the unwritten law of the hills, "let well enough alone," prevents interference. Why shouldn't the Leather Man return to the places he knew and loved so well? Didn't Farmer Clem Sorrel encounter his likeness at the Saw Mill River Cave, near Shrub Oak, New York, some weeks after the Leather Man died, and didn't it frighten him so, he ran three miles to report it? The farmer was seeking the Leather Man's money, and so realistic was his experience with the apparition that he never returned. His three-mile sprint led the New York *Daily News* on April 2, 1889, to headline the story, SAW THE LEATHER MAN'S GHOST—A Farmer Seeking for Treasure in the Dead Man's Cave Is Frightened Out of His Wits.

Among inhabitants of the Leather Man country, two time-honored statements are still to be heard: "Washington slept here," and "Leather Man ate here." Of course, one can no longer find a person who can declare that his father or grandfather invited General Washington to sleep in the guest room, but there are a number who can say with justifiable pride, "My mother fed the Leather Man."

Prior to the Civil War, the country folk of our nation were plagued with jobless men walking from town to town, some looking for work, but many asking insistently for handouts of food. Each spring a motley procession in whiskers and rags sallied forth from city refuges to the country with its well-stocked larders. Only the return of cold weather ended their foraging and the worries of rural housewives.

It was among such transients passing through the town of Harwinton, in northern Connecticut, that the Leather Man was first noticed. Unlike the others in ragged clothes, this wanderer wore leather from head to foot. This of itself was not too remarkable in an age when leather was cheap and in common use, but this outfit with its checkerboard arrangement of leather patches tied together with wide leather thongs was quite striking in staid little Harwinton.

The first time this quaint costume was seen, the residents noted it with interest, but it was quickly forgotten. To the villagers he was simply a tramp and as such belonged to a shiftless, useless lot, an object of scorn and suspicion, rather than sympathy or curiosity.

Everyone was surprised when the visitor came again a month later, and when he reappeared after another month, many questions

THE LEATHER MAN 293

were asked. No answers were forthcoming; the man in the leather suit talked to no one. Month after month he came and went over the same roads.

These repeated visits enabled the townspeople to look the Leather Man over quite thoroughly; and contemporary descriptions of his appearance agree on most points. Overlapping pieces of leather, about eight by ten inches in size, made a cumbersome coat that reached to his knees. Inside and out were voluminous pockets, in which he carried food and tools. They made the stiff and bulky coat

This is one of the few pictures of the Leather Man clear enough to reproduce. It was distributed more than sixty years ago with the "Compliments of W. P. Thoms, Waterbury, Conn." to interested visitors.

stand out from his body. His trousers, also of leather patches, looked as if they might easily stand alone.

His immense leather shoes had thick wooden soles, and turned-up toes gave a rocking-horse effect both pleasing and practical in design. Crowning the shapeless figure was what at first appeared to be a flat-topped hat with a wide visor, but closer inspection disclosed it to be a neat-fitting, patched leather cap. A close-cropped growth of beard, and shaggy brows partly obscured his bronzed and wrinkled face.

A stout walking stick with a ball handle probably provided needed support, since his leather suit, according to persons who handled it after his death, weighed sixty pounds, and his shoes ten. He also carried a huge, square pack on his back. Yet day after day he walked long distances. No one knew where he spent the time during the month which elapsed between appearances in a particular town.

Finally, in Forestville, Connecticut, Chauncey Hotchkiss became so interested in the mysterious reappearances of the Leather Man that he resolved to find out what became of him when he left Forestville. Hitching up his horse, Hotchkiss followed the Leather Man from town to town. Then, piecing together his own observations with information gained from correspondence with residents of other towns through which the Leather Man passed, Hotchkiss compiled and published a timetable of the itinerary. The article attracted a lot of attention and was widely reprinted. Odd as they seemed, the revelations were well documented.

I have read the original manuscript, which is now the prized possession of a very old woman in Forestville. According to Hotchkiss, the Leather Man made a circuit of 365 miles every thirty-four days. Approximately two thirds, or about 240 miles, were in Connecticut. This took twenty-two days, while about 120 miles in New York State required twelve days to cover.

Always traveling a clockwise course, never once reversing the direction, the Leather Man was so regular in his rounds that he could be expected at specified stopping places for particular meals, and each night "holed up" in some familiar rock shelter or cave along the way.

It was claimed that papers dropped by the Leather Man on one of his early trips revealed his identity—Jules Bourglay, of Lyons, France. This report became the subject of a great deal of speculation, and many attempts have been made through the years to verify it. Belatedly, sixty years after he died, his picture was mailed to his supposed birthplace in France, with a letter asking for information about him. But time had erased any memory of him in Lyons.

If this man was Jules Bourglay, how did he get here? How was he able at first to find his way over unfamiliar roads in a foreign country, perhaps without speaking English? One can only guess. How did he find people who would feed him and caves to shelter him? The questions are endless. Many facts about him are common knowledge, but much will always be mystery. Certainly, during the later years of his life, no man in that region was more popular, none so widely known. His admirers probably numbered in the hundreds of thousands.

However carefree the life of the Leather Man may have appeared, he bound himself to relentless schedules of time and performance for a third of a century. While no social engagements or appointments in the ordinary sense troubled him, his timetable had to be exact in order to fit into the plans of those who made special preparations for his coming. "Housewives set their clocks by him" was an oft-repeated statement that seems incredible, but it is a fact that thousands of persons on his circuit timed their activities by his movements.

Each of the thirty-four days of his circuit was Leather Man's Day somewhere. Meals were ready on his arrival, and no sooner had he eaten than he picked up his cane and heavy pack and began his five-mile stint to the house where another meal awaited him. Paying strict attention to his business of walking and eating, he had no time for dallying. It always took him the same length of time to cover the same ground.

Approaching his shelter at the end of a ten-mile day's tramp, he performed his customary chores. After lighting an inverted V of dried sticks at the front or on an inside hearth of his cave, he laid a new bed of pine or hemlock branches, made necessary repairs to the shed-like structure he had erected in the early days of his circuit, closed the opening, and stretched his weary limbs for the night.

The next morning, he gathered dry wood and stored it in the rock crevices at the back of his cave and laid another fire. About a mile down the road, a hearty breakfast always awaited him. So precise was his timing that people knew almost to the minute when to expect him. Should a meal not be set out for him at the proper time, he might never stop again. Such a calamity was to be avoided at all costs by his providers, who took special precautions to meet his deadlines. The Leather Man was never late! Or, perhaps it should be said "almost never," because the blizzard of 1888 with its twenty-foot drifts put him four days behind schedule.

Persons interested in his welfare attempted to engage the Leather Man in conversation a number of times, and once he was asked why he wore a leather suit. As on every other occasion when an inquiry was made of him, he declined to answer. He may not have understood the question, or perhaps he wished to keep the answer a secret. Some thought the leather suit was the key to the whole mystery, and that therefore the subject was a touchy one with him. Romantic speculation even went so far as to claim that he clothed himself with the cause of his ruin and was doing penance for a supposed wrong.

Less romantic people felt that the leather suit served a very practical purpose. First of all it was conspicuous, an identification for its wearer, since it looked quite unlike the clothes of other itinerants. It had greater wearing qualities, and when a part was tattered, it could be replaced with a new patch. Tanneries were almost as numerous then as gas stations are today, and scraps of discarded leather could always be found near them.

There were other reasons why his choice of leather was considered wise. His mode of living required him to be on the road every day, rain or shine. The bitter cold of New England winters would penetrate any fabric, but was bearable if body heat was retained under quarter-inch leather. Although uncomfortably hot in mid-summer, the outfit shed rain and assured the Leather Man of fairly dry garments when he retired after a drenching day.

Like the tin woodman in the *Wizard of Oz,* the suit of the Leather Man needed oiling, but for another reason. After exposure to rain and the drying influence of sun, the parched pieces, rubbing together, made a squeaking sound that announced his coming as effectively as a blare of trumpets. Mrs. Linus Mattoon of Watertown said he could be heard half a mile away, and long before he hove into sight children shouted, "Here comes the Leather Man." The noise of rubbing leather and the sight of his grotesque outfit flapping in the wind once frightened a team of horses in Woodbury. The Leather Man found it necessary to stop frequently at tanneries to have his leather suit oiled.

On another occasion, a farmer stopped at a brook by the side of the road to water his horse, and before he could check him, the animal vaulted a rail fence and landed the buggy and its occupants right side up in the nearby pasture. Upon looking around, the driver saw the Leather Man walking through the willows beside the brook.

However preoccupied the Leather Man may have been with making his rounds, he found time to take naps along his route and to read an occasional paper. No one knew whether he could read English, but people reported he scanned the shipping news and stock market quotations whenever he laid hands on a newspaper.

Mr. Frederick Barnes of Woodbury recalled a hair-raising experience that occurred when he was a small boy. A bear chained in a neighbor's yard broke loose from its tether at times and roamed the neighborhood. One afternoon when Barnes was driving cows home from pasture, the cattle became frightened and unruly. In his haste to get them herded, he forgot to watch his footing, and fell over a large, black object that moved as he touched it. Could it be the neighbor's bear?

Without pausing to investigate, he dashed up the steep hillside until he was out of breath. Only when he was at a safe distance did he dare look back. The cause of the commotion was now plainly seen. Instead of a bear, Barnes had fallen over the old Leather Man, who had been reading a newspaper oblivious of all around him. The boy watched while the Leather Man gathered up his heavy bag and cane and trudged up the swale to the Dug Way Cave.

On another occasion, in Plymouth, the Leather Man was interrupted while reading. Mrs. William Atwood of Terryville recalled a childhood incident when she was attending the school on Plymouth green. On the way one spring afternoon she decided to gather pussy willows along the road, and became so occupied with selecting choice shoots that she failed to notice where she was walking. To her surprise (and his) she stepped on the Leather Man reading by the roadside. When asked what he was reading, she said, "I believe it was *Woman's Home Companion.*"

In a northerly direction out of Woodbury, Connecticut, through a rock cut known locally as the "Dug Way," and high in the upland pasture to the west, overlooking the Nonewaug Valley, is the Leather Man's Dug Way Cave. Elbert Barnes, owner of the property, once reconstructed the shelter in front of the cave. He said that such a structure consisted of long poles resting against an overhanging ledge and supporting a thatch of dry leaves held down by a second layer of poles. An opening in the center permitted access and served as a chimney, which could be closed when necessary. Crevices in the rocks were used as shelves for storage of wood and utensils, and in one of those Barnes found the Leather Man's hatchet and pipe.

A mischievous boy once invaded this sanctuary and threw out all the firewood the Leather Man had stacked inside to dry. When his mother learned about this lack of respect for the Leather Man's property, she made him return to the cave, some two miles away, at dusk, to replace all the wood as he found it. This homely incident typifies the concern many had for the Leather Man's well-being.

Although the Dug Way Cave was the pride of Woodbury in the last half of the nineteenth century, neighboring Watertown can claim two Leather Man caves, one of which is the largest of these caves to be found.

The Watertown Black Rock Cave is the best-known Leather Man cave in western Connecticut. Located in what is now Black Rock State Park, the cave is situated at the foot of a cliff which faces south. A twenty-by-twenty-foot entrance leads to a rock-enclosed room forty feet long. The floor is strewn with the ashes of countless fires, where scouts and other campers have swapped eerie bedtime tales, and the lonely Leather Man kept his solitary vigil.

Two miles to the east, at the base of a high, hemlock-covered ledge, known as Jericho Rock, the Leather Man chose another of his sheltered cave sites. Facing the south and east, it surveys a sloping expanse of pine and hemlock that sweeps to the Naugatuck River and the Jericho Bridge of the New Haven Railroad. Robert Warner of Oakville, Connecticut, once described how the Leather Man survived the severe New England winters:

"The cave had a six to eight foot shelf of rock overhang and when enclosed was about fifteen feet long by ten feet wide and seven feet high. Poles were placed against the rock and these were covered and woven with brush and tied with bark. No rope was used. The cave was well sheltered so storms could not get in and the ground pitched out to shed the water. There were dry leaves and large pieces of wood in the back of the cave and dry twigs and leaves were arranged on the hearth ready to be lit when the Leather Man returned."

The fire was built on a large, flat stone within the cave. When a sizable bed of hot coals was formed, the Leather Man brushed it aside and lay down on the rock, which held the heat like a soapstone. The fire also warmed the overhang, creating a Dutch oven effect.

North of the Bristol reservoir, in the town of Burlington, Connecticut, is a cave known since Revolutionary times as the Tories Den. When the struggle between loyalists and patriots was at a high pitch, a number of Tories made it their hideout. A large, horizontal slab of rock rests on other rocks to form a shelter both dry and pleasantly habitable. Open to north and south, it has a good circulation of air, which dispels dampness. The snug apartment was easily heated without the annoyance of smoke because of a third opening in the ceiling. There is a good eyewitness account of the Leather Man's occupancy of this cave by the late Celester Hungerford.

"When I was about fifteen years of age, Fred Chamberlain and I went to the Den one Sunday morning and on climbing down from the cliff we came suddenly to the cave and to our surprise found the Leather Man seated at the entrance and looking right at us. The unexpected meeting and the sinister expression on the Leather Man's face, as he sat with his whole body filling the entrance to the cave frightened us so much we made a hasty exit."

A recent plane crash on Mt. Higby, between Meriden and Middletown, drew attention to a cave in traprock formerly used by the Leather Man. It wasn't until twenty-six years after the cave man died that E. W. Mildrum of East Berlin, Connecticut, reported in the Hartford *Courant* as follows:

"While tramping the mountain with companions last week, I climbed up a wooded ravine on the side of which there was an overhanging cliff, at whose base we noticed rocks piled up, having much the appearance of a miniature fort. On examination we found the remains of a fire, long extinct, and hidden among the rocks a rusty tin pail, in which were neatly coiled sections of leather taken from old boot tops, coils of leather thongs, and a wire, evidently taken from a spring bed, used, perhaps, for punching holes in leather. Here we think we have found one of the lodging places of the old Leather Man, the location of which, up to this time, has remained a secret." Situated within fifty feet of the top of the mountain, the "fort" faces east toward the Connecticut River.

One of the few photographs we have of the Leather Man was actually taken in Totoket Cave on top of Pine Rock, overlooking

Lake Saltonstall, in East Haven, Connecticut. Sixteen-year-old Joseph F. Rogers visited the Leather Man at this cave in 1885 and made a lengthy report published in the New Haven *Register*. He described the Leather Man in great detail:

"He is about the average height, though very compactly built. His thick, dark hair straggles from under the great leather cap in tangled confusion, and when closely examined the intermingled gray gives ample evidence of the burden of years which the old man has upon him. His face is swarthy and nearly obscured by a short, coarse beard and overhanging eyebrows, in whose shadows lurk his keen gray eyes with a piercing glitter. His nose and mouth are well shaped though the other parts of his face somewhat destroy their symmetry. His leather garments, from which he takes the name of Leather Man, are made up wholly of boot tops, which have from time to time, been patched and mended until scarcely a vestige of the original texture remains visible. His shoes are not unlike those worn by the peasants of Norway and Sweden, though far more cumbersome; over his shoulder he carries a leather bag and in his hand a hickory staff surmounted by a wooden ball.

"He has built the shelter I have mentioned of decayed butternut trees laid up slanting against a stick which rests upon a large boulder and the projecting limb of an ash, the hollow of the trunks being laid uppermost to carry off the rain. On the south side he has piled up brush and stones, the eastern protection being formed by the boulder itself. Within is a rude fireplace, always kept clean and ready for use, and in the nooks and crannies of the great rock his few accessories are carefully hidden and covered with leaves. A large flat stone, laid near the fireplace, is worn quite smooth, and around it are strewn a number of hemlock boughs for his bed, the stone serving in the capacity of a pillow. Without, a great quantity of wood ashes and embers have accumulated, showing that he has tenanted this secluded hut for many years. The trees in the vicinity bear no evidence of his axe as he never chops one down, but burns the old decayed wood, the hemlocks being the only exception, from which he cuts a few boughs for his bed."

The Woodbridge Cave was almost buried during construction of the Merritt Parkway. Nearby stands a memorial to the Leather Man, a small cabin made of rough cedar logs, which serves as a summer camp for its owner.

One former resident of South Wilton felt it a great privilege at the age of ten years to be able to sit down with the Leather Man in his cave, although no conversation was carried on by either of them. After a span of sixty-five years "Uncle" Dick Moriarty related this experience:

"I have been to the cave many times and sometimes when the Leather Man was there. I remember once when I had seen him pass by the school that I went to his cave that night to talk to him. He was sitting on a rock in the cave and had a good fire going. He had his shoes off and when he saw me, he grabbed them as though he thought I might take them. Of course I wouldn't take them, but he didn't know. No one ever took anything away from the Leather Man. I would try to talk to him but I couldn't understand what he said.

"The last time I saw him was about nine o'clock in the morning on March 4, 1889, twenty days before he was found dead. That was the day that President Harrison took the chair in Washington. When I heard of his death, I went to the cave and found a smooth stone, hidden on a shelf. I believe it was his whetstone, at least it puts a good edge on my jackknife."

A characteristic Leather Man rock shelter is located on the grounds of the Audubon Center in Greenwich, Connecticut. The sanctuary's director, Charles E. Mohr, first learned of its association with the Leather Man when studying reports on Indian sites. Excavations in the area had been made in 1900 by M. R. Harrington, who later was to become a noted archaeologist. The report published in the *Anthropological Bulletin of the American Museum of Natural History* contained a photograph of a Leather Man cave. Mohr recognized it immediately as the rock shelter on the Audubon property. A map confirmed the location. Harrington's diggings at the site gave proof of early Indian occupation, while scraps and thongs of leather in the upper strata of earth substantiated rumors of Leather Man tenancy.

Three miles to the north, at Armonk, in Westchester County, New York, a similar rock shelter may be found in Whippoorwill Park. The next stop in the Leather Man's itinerary was the rude hut on the Dell farm in Pleasantville, where he died March 24, 1889.

What did the Leather Man do in return for the food supplied him? We must remember he was an accepted itinerant; not a tramp. No

The Leather Man rock shelter on the Audubon Nature Center property in Greenwich, Connecticut, is the site of archaeological investigations made by M. R. Harrington in 1900. *Charles E. Mohr*

one offered the Leather Man work, and he sought none. Such diversion would have upset his exacting schedule. He relied on the friendliness of people, and few ever failed him. Some supplied a meal every thirty-four days for almost thirty years. At a time when there were few organized charities, the Leather Man provided an outlet for a personal type that could be dispensed from the doorstep or in the kitchen. Calendars were marked "Leather Man's Day," so he wouldn't be forgotten, and some housewives rose early to prepare him a meal fit for a king.

The Leather Man's silent, almost bashful, affection for children was expressed in pantomime at schoolyards and homes where he stopped. It was returned in less restrained manner. Many children followed him along the road to offer cookies and other choice items.

One little girl in Plymouth later told that she secretly wrote him messages and hid them with his food.

On the walls of the famous Honiss Oyster House, in Hartford, Connecticut, hang hundreds of pictures of merchants, writers, railroad men, politicians, actors, philanthropists, and sportsmen. Two photos of one man, hanging side by side on the left of the stairway, invite close scrutiny because of the subject's unkempt appearance. All of the persons featured here have something in common—all, that is, except this one, a man dressed in a strange leather outfit. It has been said of Jules Bourglay that he climbed the ladder of success downward to fame.

A full explanation of his popularity may never be given, but its existence cannot be denied. A short time ago, near the city of Ossining, New York, nearly one hundred people gathered in Sparta Cemetery to dedicate a plaque at the Leather Man's grave. For sixty-four years no marker had been set there for lack of positive identification. Was this actually Jules Bourglay? No one knew for certain, but at least the Leather Man was buried there, and the two names had been associated for so long that they had become synonymous.

A shiny brass plaque attached to a low granite stone makes this brief declaration:

FINAL RESTING PLACE OF
J U L E S   B O U R G L A Y
OF LYONS, FRANCE
"The Leather Man"
Who regularly walked a 365 mile route through Westchester and Connecticut from the Connecticut River to the Hudson living in caves in the years
1858-1889

The Cave in Rock Murderers

George F. Jackson is a native Hoosier who has studied and photographed Indiana caves since he was a teen-ager. He has no clear recollection of how he became interested in Cave in Rock, which is in Illinois, except that its gruesome history had intrigued him even as a boy. Today at forty-seven, he believes he has collected more information about this cave than anyone else.

GEORGE F. JACKSON

As caves go, it isn't much. It is located on the north bank of the Ohio River in what is now southern Illinois. The rocky cliffs along the river banks are high and sheer, and from a distance they appear to be smooth. Prominent on the face of one of these limestone bluffs is the huge arch of the cave opening, seeming almost too regular to be entirely natural. But it is, and although the giant forests around it have disappeared, it must look today very much as it did in 1729 when first seen by early French explorers. Peaceful enough then and now, but Cave in Rock has had a bloody and violent history.

One night in 1799, a blazing fire burned at the mouth of the cave, which yawned like an immense crypt, silent and foreboding. High above, and for long distances on either side, the dirty grey stone of the bluff was thrown into sharp relief by the leaping flames—flames that were reflected in the dark waters of the Ohio, a few feet below.

This was no ordinary campfire. This was a victory celebration, a roistering, dirty feast likely to end in a dozen free-for-alls. Suddenly there was a scrambling and breaking of branches on the bluff a hundred feet directly above the cave. Startled into immobility, the dozen or so men around the fire stared up at the cliff. With wild shrieks of terror, a naked rider, bound to a horse, came hurtling down directly upon the blaze. From above, great peals of laughter rang out.

These were the most desperate and bloody murderers and thieves in the history of the Middle West, men who terrorized most of Kentucky and Tennessee and parts of the neighboring states. And in 1799, Cave in Rock was their rendezvous. Seldom has nature con-

Entrance of Cave in Rock, as seen from the waters' edge of the Ohio River. *George F. Jackson*

structed such an impregnable pirate stronghold. Extending back into the hillside for some distance, it offered ample protection from the elements and from officers of the law for a sizable band of men. At the back end, a natural crack extended upward to the surface. The pirates normally used this chimney only as an outlet for smoke from their fires, but it could, if necessary, serve as an escape route.

When the cave first became a rendezvous for desperate criminals, no one knows, but the first notable use of it was by Captain Samuel Mason, an officer of the Continental Army, member of a distinguished family, and hero of many Indian battles and raids. Few outlaws have had such an honorable background. What caused Mason, almost in middle age, to change from an honest man to a shrewd and resourceful highwayman, no one has recorded.

He arrived at Cave in Rock around 1797 and immediately went to work, with a daring plan of operation. Setting up living quarters within the cave, he posted a large sign on the river bank, reading LIQUOR VAULT AND HOUSE FOR ENTERTAINMENT.

His carefully worked-out scheme was based on the theory that river boat personnel, seeing the sign, would disembark for rest and entertainment. Later, while being "entertained" by members of Mason's band, they could be dispatched with ease and the boats looted at will.

The plan worked beautifully. If any of his patrons came to suspect that the sign was a ruse, they never lived to warn others. Although strange tales of vanished craft and unusual happenings on the lower Ohio drifted back to civilized points, it was a long time before men began to be suspicious of the cave in the river rocks.

If it seems strange that the crews of the early river boats could be enticed to tie up at such a spot, consider the situation and the times. The only real way to the West and South was via the Ohio to the Mississippi. Any traveler or shipper wishing to visit or send goods to any point west or south of the present Middle West was forced to travel the stream. It was a place of dangerous currents, treacherous, shifting sand bars, islands, and rapids. No man knew it thoroughly. Some of the more dangerous spots were marked on crude maps, but at many points it was the custom—the necessity even—to engage local guides to pilot boats through the channels and rapids. One extremely dangerous place was the Hurricane Island rapids, just below the cave.

It may be that some boats stopped at Mason's House of Entertainment for a brief rest before entering the rapids, others, after passing through them, for relaxation; still others, because they were manned by tough crews, certain that they could "lick any man alive," may have welcomed a hazardous encounter. It was a member of one of those river boat crews who boasted: "I am a man. I am a horse. I am an alligator. I can whip any man in all Kentucky, by God!"

At any rate, there is no doubt that both crew and passengers ordinarily looked forward to any stopover. The last important stopping-place upstream had been the "Falls of the Ohio," later Louisville, Kentucky, and, after this distance, such an intriguing spot as Cave in Rock was an irresistible magnet.

But there were some river men who did not stop. For them, Mason developed another plan. When a boat was seen approaching, he stationed a man at Battery Rocks, ten miles above the cave. This man would hail the craft and offer to guide it through the difficult sand bars and the Hurricane Island rapids below. Many crews took advantage of this "service," but some declined. To take care of the latter, Mason had still another man ready a few miles farther downstream. If the first man failed to get aboard the craft, the second was likely to do so. Once aboard, the "pilots" had an easy time running the boats aground, either in front of Cave in Rock or at the head of Hurricane Rapids. Mason's well-armed desperados would then leap upon the bewildered crews, and if all were not dispatched with ease, there is no record of it today.

Not all boats were grounded and looted; those with cargoes of little value were actually guided through the rapids for legitimate pay. It was a foolproof setup.

For some time, Mason's "business" prospered, but eventually rumors began to reach him of a group being organized up-river in Pittsburgh, to ferret out the mystery of vanished river craft. He began to look for another field of operation, and soon found it in the Natchez Trace, the Wilderness Road that ran from Natchez, Mississippi, to Nashville, Tennessee. Many of the travelers along this old and narrow Indian trail were merchants who had sold the goods they had brought down the rivers and were returning northward with the proceeds of the sales. It was a rich field for a reckless highwayman.

Mason robbed and murdered for several years along the Trace and the Mississippi River, but eventually he was double-crossed and slain by an even more despicable killer, "Little" Harpe, about whom we shall hear more later. Harpe, although himself wanted for murder by the authorities, had the temerity to decapitate Mason, encase his head in clay to preserve it, and take it to the seat of the territorial government for a reward.

Another unscrupulous character operating along the dangerous stretches of the Ohio River near Cave in Rock during this same period was a Colonel Fluger, who used an entirely different approach from Mason's. Although the "Colonel's" exact military title and the correct spelling of his name (it is given by various historians as Plug, Pflueger, Flueger, and Fluger) are in doubt, there is no doubt of his exploits as a boat-wrecker and murderer.

He would wait until a boat tied up along the banks, then, slipping aboard it, would bore holes in the bottom. When the boat began to sink, he and his confederates would murder the passengers and crew and take whatever goods they could use for profit. The Colonel's end was quite appropriate: boarding one well-laden craft, he stealthily made his way below decks and unplugged the seams. The water rushed in so violently and unexpectedly that when the boat sank, Colonel Fluger was the first and only fatality.

About the time Mason was shifting operations to the Natchez Trace, the two Harpe brothers, the most bloodthirsty outlaws ever to infest the Ohio River, were coming up through central Kentucky and Tennessee. Their recorded history is appalling, but their total history must be even worse, when you consider that most of their victims never lived to give an account of what happened.

They came, originally, from North Carolina. Micajah, known as "Big" Harpe, was born about 1768; Wiley, known as "Little" Harpe, about 1770. Their early history is little known, but somewhere around 1795 they left North Carolina with two women, both of whom were claimed by Big Harpe as wives.

For several years, they roamed central Tennessee, spending some time with "outlaw" Indians who were committing outrages against their own people as well as the white men. Not only did the Harpes help the Indians, they even added some embellishments of their own to the Indians' brutal practices.

In their progress from Tennessee to Kentucky they left a trail of murders. They never made any attempt to hide that trail or to disguise themselves, once they came upon an unwary traveler. When they met large parties or were in settlements they obeyed what they thought to be common sense and refrained from killings in plain view of others.

For a time they stayed in Knoxville, then a roaring town at the gateway of the Wilderness Road to the West. It was here that Little Harpe gained enough semblance of respectability to court and marry the daughter of a well-known local preacher. But Harpe's contempt for law and order soon reasserted itself, and before long he and his brother were stealing livestock, drinking and carousing, and getting into minor trouble in town. Several houses and barns were set afire and burned to the ground. No motive could be discovered by the authorities, but suspicion turned toward the Harpes when it seemed obvious that the properties had been destroyed from sheer wantonness.

Soon afterwards, Edward Tiel, owner of a stable of fine, blooded horses, found that several of his stallions were missing. Instantly he thought of the Harpes. Summoning several hired hands and neighbors, he rode over to their cabin. It was deserted, but there was ample evidence that horses had recently been about. This was enough for Tiel. Followed by his posse, he galloped into the woods, determined to bring the thieves to justice. The Harpes had made no effort to hide their trail, and it was easy for Tiel and his men to track them into the Cumberland Mountains. There, deep in the heavy forest, they came upon the two Harpes, driving the stolen horses before them. As inconsistent in their behavior as ever, they made no resistance but docilely permitted themselves to be led back toward Knoxville.

Five miles along the way, however, they somehow threw off their bonds, sprang into the forest, and disappeared in the underbrush. Despite an intensive search, they were not seen again. It was as if the ground had swallowed them.

Thus the Harpes left Knoxville.

A few days later, a body was found in the Holstein River. It had been crudely cut open, the intestines removed and the space filled with stones, usually an effective method of disposing of unwanted corpses. This time, however, it had been hurriedly done, and the body had floated to the surface. Shortly after this gruesome discovery, a man named Hughes, who ran a disreputable tavern in the vicinity, came up with a story that the corpse was that of a traveler named Johnson, and that the Harpes had killed him in the tavern. His tale was substantiated by two of his wife's brothers, named Metcalfe. All were known to be associated with the Harpes, so all were arrested and placed in jail. Although vigilantes later chased Hughes out of the country and burned down his tavern, both he and the Metcalfes were acquitted at their trial. Of Johnson, no one seems to know more than that he was a traveler and that he was the first known of the many victims whose bodies the Harpes tried to dispose of by this method.

The Harpes had not forgotten their women. Moving westward along the Wilderness Road, they met the women somewhere near the Tennessee-Kentucky border. One of the mysteries about the Harpes is that the three women who accompanied them—all of whom were reported to have come from cultivated families—stayed with the outlaws through atrocity after atrocity, peril, hunger, and the many other hardships of the frontier.

Moving northwestward through Kentucky, they killed a peddler named Payton and took his goods; tomahawked and shot from behind two men from Maryland, Paca and Bates; and near Crab Orchard killed a wealthy traveler from Virginia, who had asked their company through an exceptionally dangerous section of the trail.

After this last murder, they were apprehended by a posse from Stanford. Eventually all five were placed in the county jail at Danville. Despite unusual precautions, such as extra locks on the doors, "two horse locks to chain the men's feet to the ground," and the hiring of four extra men, two of whom were to guard them at all times, the Harpes escaped, leaving their "wives" behind. This may have been because each was expecting a child within a short time.

The babies were born in jail, and when the women's trial came up, they were acquitted and even given food, clothing, and a horse by the chivalrous citizens of Danville. Instead of heading eastward, however, as they were expected to do, they, with their three children, went directly to Green River, secured a boat (probably by trading the horse for it), and paddled their way downstream to a meeting with their husbands. Although there is no reliable record of just where the meeting took place, some old accounts have it as occurring in Mammoth Cave or one of the other large Kentucky caverns which were known even then.

Wherever the meeting, the whole Harpe party made its way to the Ohio River Valley and Cave in Rock. How many murders they committed on this trek will never be known, but there are definite records of at least one, and possibly several more.

A large group of outlaws had been living at the cave prior to the Harpes' arrival. One historian says that, since most of this group had been chased out of the so-called "law-abiding" communities—which themselves were notoriously wild and tough—it can be imagined that the band at Cave in Rock was completely lawless and depraved. Yet even these cutthroats found the Harpes too much to stomach.

According to one story, the Harpes had hardly arrived when a flatboat came down the river and landed not far above the cave, at a place known as Cedar Point. The passengers, not knowing they were near a nest of outlaws, had gone ashore and were strolling along the river banks. Among them were a young man and his bride-to-be. These two strolled to the top of a bluff and sat down, looking out across the river. The Harpes, who had been watching the scene, sneaked up behind the two lovers and pushed them over the high cliff. To the Harpes this was a joke, and they returned to the cave laughing uproariously about the trick they had played. Apparently it did not have quite the same effect on the others of the band. They did not like it, and they told the brothers so.

Shortly afterwards, there occurred the incident with which we began this story. A boat had been captured and most of its passengers slain. Not all of the men, however, had been killed; some were taken captives and tied up, while the outlaws debated what to do with them. The Harpes quietly took one of these captives, stripped him, tied him to the back of a horse, and led the animal to the top of the cliff, one hundred feet above the cave. Below, the rest of the band was gathered around a great fire, talking. Suddenly the Harpes drove horse and bound rider over the cliff's edge . . .

The other outlaws found this feat a little too much for even their cruel appetites and actually drove the Harpes and their party out of camp.

The Harpes committed twenty-two more known murders (including those of several women and children) while wandering in central Kentucky. It must have been during this period that Big Harpe took one of his own children and dashed its head against a convenient rock, because its crying bothered him. Just before Big Harpe died, he said this was the only one of his acts he regretted.

He was finally killed and beheaded near Robertson's Lick, Kentucky, by a man whose wife and baby he had slain in their sleep. His head was mounted on a pole at a prominent crossroads and for a long time served as a revolting warning to other bandits. Even today, the spot is known as Harpe's Head.

As for Little Harpe, he made his way to the Natchez Trace, changed his name to Setton, and, as we have seen, double-crossed his fellow outlaw, Mason, severed his head, and took it to court, expecting to collect a substantial reward. Just as he was about to collect—and probably receive the plaudits of the populace—a Kentuckian recognized the horse tied outside as one that had been stolen from him some days earlier. This was the beginning of the end for Little Harpe. He was arrested, tried for highway robbery, and hanged in Greenville, Mississippi Territory.

Although Mason and the Harpes were the most notorious of the Cave in Rock criminals, there were many lesser ones who, through the years, used the spot as a hideout or for their headquarters. It was used by at least two groups of counterfeiters, but their activities are not so well known as are those of the larger bands who preyed on the river traffic.

Several early writers mention the place as being the workshop of counterfeiters, and a number of molds and dies have been unearthed in and near the entrance. One of the counterfeiters is said to have taken two women who befriended him to the cave and, after blindfolding them, to have led them up into a huge "secret room" where he showed them "large quantities of counterfeit silver and gold coins in boxes and chests." One still hears the old rumor that this "upper room" of the cavern actually exists, and that there is a great store of good and bad money and piles of jewels and other valuables. The fact that geologically such a cavern is impossible does not dampen the treasure-seekers' spirits.

Probably the last real outlaws who used the cave were the Ford's Ferry band, a mysterious group about which much fiction, but little real fact, is available. When the last member of the band, Henry C. Shouse, was hanged for murder on June 9, 1834, it marked the end of the Cave in Rock outlaws.

Today, Cave in Rock is the chief attraction of the state park so named, and although the grounds have been cleared of underbrush and undoubtedly appear less wild than when first seen by pioneers, the great cave still looks much the same as when the French explorers glimpsed it from the blue waters of *La Belle Rivière*. It stands above the peaceful river, and there is no outward sign of its bloody past and no visible trace of the violent men who contributed such an unsavory footnote to American history.

Mark Twain Cave

Mark Twain Cave is owned by Archie K. Cameron, a resident of Hannibal, Missouri, and a member of the National Speleological Society. Today the cave is a tourist attraction for the thousands of visitors who come to Hannibal to see the points of interest made famous by Samuel Clemens' writings.

In Clemens' day it seems to have been almost as busy, for he once wrote, "I think my mother was never in the cave in her life; but everybody else went there. . . . It was miles in extent and was a tangled wilderness of lofty clefts and passages. It was an easy place to get lost in; anybody could do it."

Howard N. Sloane

Tom Sawyer and *Huckleberry Finn* are based, to a large extent, on Samuel Clemens' own boyhood experiences. Many of those experiences involved a cave about two miles south of Hannibal, Missouri, where Clemens lived as a boy.

Originally it was called Simms Cave, after Jack Simms, the hunter who first discovered it in 1819. Later, it was named Saltpeter Cave and then McDowell Cave, the name by which young Clemens knew it.

The deep impression that McDowell Cave made on the mind of Sam Clemens was to stay with him throughout his life. In almost everything he wrote there was at least a reference to the cave. Even while traveling through Europe, he was unable to cast out his boyhood memories of its dark passageways. In *Innocents Abroad,* he writes, "Cave is a good word when speaking of Genoa under the stars; when we have been prowling at midnight through the gloomy crevices they call streets, where no footfalls but ours were echoing, where only ourselves were abroad, and lights appeared only at long intervals and at a distance, and mysteriously disappeared again, and

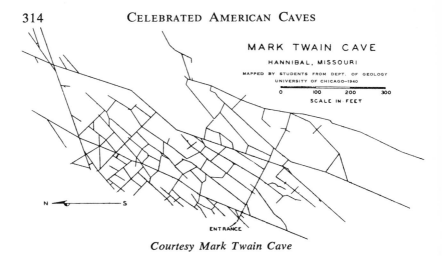

Courtesy Mark Twain Cave

the houses at our elbows seemed to stretch upward farther than ever toward the heavens. The memory of a cave I used to know at home was always in my mind, with its lofty passages, its silence and solitude, its shrouding gloom, its sepulchral echoes, its flitting lights, and more than all, its sudden revelation of branching crevices and corridors where we least expected them."

It is no wonder the cave was easily remembered, for during the 1840's Dr. E. D. McDowell, an eccentric St. Louis surgeon, placed the corpse of a fourteen-year-old girl, said by some to have been his daughter, in a glass and copper cylinder. This he filled with alcohol, and suspended from a rail which bridged a narrow passage in the cave. McDowell claimed this was an experiment to see if the limestone cave would "petrify" the cadaver. It was said that "loafers and rowdies" would drag the small body up by the hair to view the dead face. Undoubtedly, the corpse served its purpose as a tourist attraction. McDowell's eccentricities extended to using the cave for the storage of cannon and small arms "for the invasion of Mexico."

The entrance to the cave is about half a mile from the Mississippi River, at the foot of a wooded limestone hill known as Cave Hollow. Many of the steamboat excursions along the Mississippi from Quincy, Illinois, stopped at this cave in the late 1800's. Horse-drawn wagonettes hauled the excursion visitors unwilling or unable to walk along the path to the cave.

Clemens, in his autobiography, tells how "General Gaines," the town drunkard, got lost in McDowell Cave for a week. According to the story, he finally pushed his handkerchief out of a hole in the hilltop "several miles" down the river from the cave's mouth; it was seen and Gaines was rescued. Clemens says, "There is nothing the matter with his statistics except the handkerchief. I knew him for years and he didn't have any, but it could have been his nose; that would attract attention."

Sam played hooky from school one day, and visited the cave. Returning home, he told his mother, "There's something in my coat pocket for you." She hardly expected to find a pocketful of bats. As Clemens says in his autobiography, "She would put her hand in but she always took it out again, herself; I didn't have to tell her. It was remarkable the way she couldn't learn to like private bats. The more experience she had, the more she couldn't change her views." Young Sam Clemens did not share his mother's opinions. He thought a bat was "as friendly a bird as there is—beautiful, soft and silky—pleasant to the touch and grateful for caressings if offered in the right spirit."

Today McDowell Cave is named Mark Twain Cave. The bats have vacated and moved to other quarters since the visitor's circuit is now electrically lighted, but in Clemens' writings, the bats play an important part. In *The Adventures of Tom Sawyer,* "Injun Joe," the half-breed villain, starves to death in the cave. Clemens says that in real life he *would* have starved to death if the bats had run short, and offers as an apology that he invented the incident in the "interests of art; it never happened."

Many tales of buried treasure in the cave originated in the Gold Rush of 1849. Sam Clemens himself spent hours digging for gold in the cave entrance, and the story of the robbers in *Tom Sawyer,* and the scene of Tom and Becky Thatcher lost in the cave, were both inspired by these legends.

Sam Clemens himself, "along with a lady," was actually lost in the cave. They were finally rescued after their last candle had burned down to almost nothing. The "lady" was really little Annie Laurie Hawkins. Sam and his gang made a habit of playing in the cave, and leaving trails of kite strings to mark their progress. Their constant ventures into the cave bolstered their confidence until they felt they knew their way well. On this particular day, there had been a picnic party, which was followed by a visit to the cave. Somehow, the members of the party worked their way into passages where they had never been before. Annie Laurie cried, frightened by the weird formations and the damp coolness of the air, but Sam was wise enough to conserve their candles carefully. Ultimately, they saw the lights of the rescue party flickering in the distance.

No better description of the cave can be given than the one written by Clemens himself in *Tom Sawyer*.

" . . . they wandered down a sinuous avenue holding their candles aloft and reading the tangled webwork of names, dates, post-office addresses, and mottoes with which the rocky walls had been frescoed (in candle-smoke). Still drifting along and talking, they scarcely noticed that they were now in a part of the cave whose walls were not frescoed. They smoked their own names under an overhanging shelf and moved on. Presently, they came to a place where a little stream of water, trickling over a ledge and carrying a limestone sediment with it, had, in the slow-dragging ages, formed a laced and ruffled Niagara in gleaming and imperishable stone. Tom squeezed his small body behind it in order to illuminate it for Becky's gratification. He found that it curtained a sort of steep natural stairway which was inclosed between narrow walls, and at once the ambition to be a discoverer seized him. Becky responded to his call, and they made a smokemark for future guidance, and started upon their quest. They wound this way and that, far down into the secret depths of the cave, made another mark, and branched off in search of novelties to tell the upper world about. In one place they found a spacious cavern, from whose ceiling suspended a multitude of shining stalactites of the length and circumference of a man's leg; they walked all about it, wondering and admiring, and presently left it by one of the numerous passages that opened into it. This shortly brought them to a bewitching spring, whose basin was incrusted with a frostwork of glittering crystals; it was in the midst of a cavern whose walls were supported by many fantastic pillars which had been formed by the joining of great stalactites and stalagmites together, the result of the ceaseless water-drip of centuries. Under the roof vast knots of bats had packed themselves together, thousands in a bunch; the lights disturbed the creatures, and they came flocking down by hundreds, squeaking and darting furiously at the candles."

Lost Footprints

George F. Jackson knows more about Wyandotte Cave than any other person alive. Not only did he at one time work as a cave guide in this commercial cave, but in 1953 he wrote a book about it. He has explored every nook and cranny of the cave that it is humanly possible to get into. One of the earliest members of the National Speleological Society, Jackson has been active in its affairs for many years and has had well over a hundred articles published on caves, most of them illustrated with his own photographs. In this story of Wyandotte's prehistoric explorers, Jackson raises some perplexing questions. Many of the answers may never be revealed, but Jackson's efforts to find them have uncovered many clues.

The first written account of this cave appeared in 1806. The author was William Henry Harrison, then governor of the Indiana Territory, later the ninth President of the United States.

GEORGE F. JACKSON

There is no actual record of the first white men to enter Wyandotte Cave. The earliest date historians associate with the cave is 1798, but it was known to Indians many hundreds of years before civilized man reached what is now Crawford County, Indiana, not far from the Ohio River.

The cave takes its name from the Wyandottes, a branch of the powerful Iroquois tribe. When white men began to make conquests of land in the area, they found wandering bands of Shawnee and Cherokee Indians from Kentucky using it as a hunting ground, while Lenape or Delawares from the north-central part of the state made occasional forays for food. It may have been these tribes which left the broken flint chips, the huge piles of imperfect arrowheads, the burnt torches, and the charred remains of sedge grass and hickory bark in the cave. It could be their peculiar implements which were found in the hills and fields near the entrance.

Whether or not a race of people predating the modern Indian also traveled through the cave, searching perhaps for flint and other stones, is open to conjecture. Some authorities believe so. Certainly there is ample evidence that the cave was visited long before white men ever came to this part of America.

In America, the search for prehistoric relics in caves continues incessantly, but the surface has only been scratched. This is especially true at Wyandotte, where there has been only casual archaeological work since the beginning of the saltpeter mining era there early in the nineteenth century. However, even the small bits unearthed have revealed many curious things.

In the dry portion of Wyandotte there is ample evidence that some early race frequently visited, lived in, or traveled through the cavern. Bits of burned hickory bark, pieces of reed, poles, grass, even fragments of woven materials, are found, sometimes under rock, beneath deposits that have taken hundreds of years to accumulate.

Ever since these traces of aboriginal passage were first noticed, speleologists have been asking, "Did the early cave visitors have special reasons for entering the cave besides simply looking for flint or shelter?" With the crude torches they used, they could not have entered simply for the sake of exploration. Were they searching for something, and did they actually find and carry away what they were looking for? No one knows for certain.

When white men first ventured into the Old Cave Route of Wyandotte, they found not only the artifacts mentioned above, but at the end of a passage, deep down in the cave, a tremendous stalagmite, known today as the Pillar of the Constitution. At its base was a huge excavation. For some time it was believed that this had been created by the saltpeter miners around the year 1812. In 1877, however, Professor John Collett found several glacial boulders, not native to the section near the Pillar. He stated in 1878 in the *Indiana Geological Survey* that, "from indications, such as wear and bruises, they had been used as hammers, or grinding pestles and proved conclusively that that part of the cave had been visited, if not occupied by men of the 'Stone Age.'"

Soon afterwards, also in 1878, Dr. Horace C. Hovey claimed that the excavation in the Pillar had been made by Indians, and from that day to the present, evidences have occasionally been found that some race of men, before the advent of the white man, had chiseled away a great part of the base of this huge stalacto-stalagmite. In all, more than a thousand cubic feet of the formation have been removed, although the hole is insignificant when compared with the entire

The Pillar of the Constitution in Wyandotte Cave, Indiana, one of the largest stalagmites known. It is thirty-five feet high and seventy-five feet in circumference. The man is standing in the area where more than a thousand cubic feet of the formation were chiseled away by Indians. *George F. Jackson*

column. Much of the broken calcium carbonate from this hole is still in the cavern, scattered about the room in which the formation stands.

A complete search for more information about these early miners was conducted by W. S. Blatchley in 1896, in the course of considerable archaeological work in the room where the Pillar stands. His measurements indicated that a space eight feet long, six feet wide, and five feet deep had been quarried from the column above the floor level. At the base of the Pillar he dug a trench through the debris to the solid stalagmite beneath. A perpendicular section of the trench revealed that the debris consisted of ashes, charcoal, and rectangular flakes of broken stalagmite, built up over a considerable period of years.

In the trench, also, were found six quartzose boulders, weighing from three to six pounds, and five different deer horns which had evidently been used as wedges. Since then, several other hammer stones have been unearthed in the same room, some with depressions worn in them by the fingers of the old quarrymen.

No one has been able to ascertain for what enterprise this material was used by the old miners. Very few Indian implements of stalagmitic rock have ever been found, since this rock soon decomposes when left to the elements. Yet, the material must have been of great value to those who dug it out, for it was secured only after a tremendous amount of work, and it seems reasonable to assume that some of it would have been found in sheltered spots outside the cave. What, then, became of it? This is, perhaps, the most puzzling question in connection with the first explorers of Wyandotte Cave.

It may be argued that the Indians visited this part of the cave for flint, as they did the New Cave section, but the Old Cave lies a hundred or more feet above the flint strata, and no flint is exposed anywhere in its passageways. This indicates strongly that it was only for the pieces of stalagmitic formation that they climbed over the rough, hilly, and pit-filled reaches of this branch of Wyandotte Cave.

In 1850, when a way into what is now called the Short Route was dug out, other traces of earlier explorers were uncovered. The walls and ceilings of some rooms and passageways were blackened by smoke from fires and torches, and great quantities of hickory bark, poles of sassafras, pawpaw, and other soft woods easily cut with stone axes were found. Close examination of some of the poles will soon convince even the most skeptical that they were cut by dull stone tools.

Many of these poles lie immovable under masses of stone that must have dropped from the ceilings hundreds of years ago, as formations and flowstone have already grown on top of the rocks. In one place, called the Animal Pit, the author has carefully examined a stick that rests beneath a huge stone and is convinced that it was there long before the rock dropped from above. If this great rock fell at the same time as others not far away, and if the word of scientists as to the age of the formations on similar rocks is correct, the stick would be of a staggering age. Since this part of the cave has been traveled for long over one hundred years, and no information that *any* rock has fallen in that length of time can be found, there is indeed cause for speculation. It is known that the stick has been there for over 150 years, but for how many years before that?

The New Cave part of Wyandotte seems to have been quarried for flint, probably by modern Indians. Flint ledges, several inches wide, are exposed in numerous places, and in one spot a deep ledge has been carved out across the cavern. The floor near these places is literally covered with broken pieces of the once highly prized material. Neither implements nor weapons were made in the cave; the Indians evidently preferred to remove the flint to the better light of day, or to the village workshops for this delicate process. Archaeologists have located many large flint workshops in Harrison and Crawford counties, Indiana, and undoubtedly some of the flint which was shaped into the necessary tools and weapons came from Wyandotte.

Most of the broken pieces of flint in the cave are rectangular in shape and average about four inches by two and one-half inches in size. Few of these show marks of tools of any sort, but they were broken from the head-high ledge by some artificial means.

Near each quarry are smoke-blackened walls and innumerable pieces of wood and sticks which show traces of having been in a fire. Throughout the flint beds are found pieces of grapevine, sedge grass, reeds, and other vegetable matter. Careful search has proved that the bundles of reeds or bark used for torches were tied together with grapevine ropes. Most of the bundles seem to have been about four inches in diameter, judging by the size of the loops of vine. No complete loops larger than this have been found, but sections of what were longer ropes have been dug up here and there. The larger bundles were probably a reserve supply of fuel and torches. The almost unbelievable amount of hickory bark, easily stripped from the shagbark hickory trees, shows that this must have been a valued fuel and torch wood.

Intentionally shaped tools have never been found in Wyandotte, although several arrowheads and other flint tools have been un-

Two of the grapevine knots found in Wyandotte Cave, Indiana. They were used by Indians to tie bundles of reed and bark used for torches. *George F. Jackson*

earthed in small caves in nearby territory, usually just inside the entrances. These are of rather late Indian workmanship, however, and may not have come from the same tribes which did the mining. Considering the number of years that the cave was "worked," it seems inconceivable that broken tools, at least, were not dropped or discarded here and there. Perhaps there are many such implements among the piles of broken rocks on the floors, and later investigations may bring them to light.

Some quarrying of stalagmitic material was also done in the New Cave, but in much smaller quantities than in the Old Cave Route. On the crest of Monument Mountain, a formation off to one side of the regular path, there are a number of broken pieces of spar, and on a large stalagmite, four feet high and ten feet in circumference, the marks of crude hammers and chisels show plainly.

But this part of the cave seems to have been used chiefly for the quarrying of flint, which was broken from the walls anywhere it could be reached. In a dry and dusty room some thirty feet from the largest flint quarry, Mr. Charles J. Rothrock and the author excavated a great quantity of small bird bones, reeds, grass, charred wood, and pieces of stalactites, indicating that this may have been a gathering place for the miners while they waited their turn to break the flint from the walls. Near this spot were found pieces of crude rope, woven fabric, an immense grass bed, and the mummified foot of a deer.

Late in 1850 the continuation of what is now the Long Route was made possible by enlarging a small hole through an almost solid wall of flowstone. It has been estimated that the time required for this sheet of calcium carbonate to form was from 2,000 to 3,000 years. The explorers found no opening through which a man could make his way, yet, when the breakthrough was completed, they discovered moccasin tracks scattered all over the floors *on the other side*.

The tracks were particularly plain in Morton's Marble Hall and the Crawfish Spring Branch. The footprints wandered from side to side of these avenues, but seemed to point in one direction only— inward. The prints continued for some distance in both passages but finally disappeared among the rocks on the floor.

At first it was believed that the tracks had been made before the flowstone wall was closed by natural processes, although several thousand years is a long, long time. This theory may be as correct as any other; however, a later theory suggests that the explorers entered the cave through another entrance. If so, this entrance was purposely or accidentally well concealed, for not the slightest trace of it has ever been found, despite repeated searches.

An interesting point about the Indian tracks is that the clay on the floor is now, and was in 1850, of a hardness that can be marred only by a sharp heel kick. Moccasin tracks certainly would not show up on earth of that consistency. These Indian footprints must have been made before the mud had dried to its present state; hence, the tracks must be of great age.

These are the known facts about the possible prehistoric explorers of Wyandotte Cave. All else is speculation. The story is incomplete, and the archaeological possibilities are unlimited. Perhaps in the not too distant future some aspiring speleologist may coax from the depths of this cave more secrets of its past.

Index

(Figures in italics refer to illustrations.)

Abbey's Ferry, Calif., 201
Academy of Natural Sciences of Philadelphia, 41, 183
Accidents: air cut off, 101; falling rocks, 125, 129, 163, 178, 182; falls, 32, 77, 262, 281-82; fatal, 77, 125, 129, 170. *See also* Explosion; Fatigue
Ackerly, Ernest, 263
Adams, L. S., 216-21
Adaptations. *See* Fauna
Adventures of Tom Sawyer, The 313, 315-6
Aguilar, Ramon, 259, 260, 261, 266
Air circulation. *See* Circulation
Aitkin Cave, Pa., 230, *231*, 232
Alabama, 16, 17, 135
Albuquerque, N. M., 186
Alhinc, French narrator, 201
Allegheny Mountains, 138, 141, 175, 226-27
Alligator, 175
Alta California, 200-201
Amphipods, 285, 286
Andy's Ridge, Md., 173, 175-76
Anastamoses, 6. *See also* Development
Animal Pathology Laboratory, Mexico City, 54
"Animal Room" in Marvel Cave, Mo., 271, 273
Anoptichthys antrobius, 268
A. hubbsi, 268
A. jordani. See Blindfish, Mexico

Ansley, Jack, 228-29
Antelope, 175
Anthodite, 133
Anthropological excavations: in eastern bone caves, 175-84; in Moaning Cave, 203-204; in Sandia Cave, 185-92
Antibiotics, 4, 116
Antiguo Moreles, Mexico, 57
Antrozous, 219
Appalachian Mountains, 8, 16-17, 130-141, 173
Aquatic life. *See* Fauna
Aragonite, *151, 155*
Archeological investigations:
  California bone caves, 194-205; age of bones, 203; bibliographic research, 194-202; bone piles, 195; cave names, 194; Chehalumche, 195; mineral incrustations, 203-205; prison, 197-98; trap-like entrances, 194-95, 199-200
  Cuba: petroglyphs, 252; "snake serpent", 244, 247; Zemí (Idol), 244, 252-55, *252, 253*
  Durham Cave, Pa., 183
  Jones Quarry Cave, W. Va., 181
  Leather Man Cave, Conn., 301, *302*
  Mammoth Cave, Ky., Indian mummy (Lost John), 124-29, *127, 128*
  Sandia Cave, N. M., 185-92
    artifacts, 186-88; Carbon-14 dating, 185, 192; European counterpart, 191; Folsom Man, 185, 188, 190-91; glacial advances and climate, 191; Pueblo culture, 187, 189; Sandia Man, 186, 191-92; stratigraphic evidence, 188-91, *190*

Wyandotte Cave, Ind., 318-20; stalagmitic quarry, 318-19, *319*

Arizona, 5

Arkansas, 17, 18, 269, 288

Armbuster, Raymond, 175

Armonk, N. Y. 301

Army Medical Service, 58

Arnold, Roy, 104

Art. *See* Carvings, Pictographs, Petroglyphs

Artibeus, 48

Artifacts: bark, 321; carvings, 252-53, *252, 253;* Ciboney Indians, 244-245; Folsom points, 191; grapevine rope, 321, *322;* Mammoth Cave, 121, 126-27; stalagmitic covering, 190-91, *190,* 203; Taino Indians, 244; mentioned, *14,* 124, 183, 186, 195, 249, 317-18, 320, 323

Astyanax fasciatus mexicanus, 265, 258-68; adaptation, 268; derivatives, 268; hormonal imbalance, 268; predation on *Anoptichthys,* 267. *See also* Fish

Audubon Center, Conn., 301, *302*

Audubon, John James, 206, 223

"Bacon rind", 34

Bacteria, 148

Bailey, Vernon, 152

Baker, Isaac W., 194-201

Baltimore and Ohio Railroad, 131

Barden, Robert, 279

Barnes, Elbert, 297

Barnes, Frederick, 297

Bat Cave, Okla, 285-87, *287*

Bat(s):
  banding, 153, 229-39, *229, 231;* Clemens on, 315; colonies, in East, *217,* 225, *225;* in Midwest, 285; in Southwest, *207, 210,* 216, 221
  flights, Devil's Sinkhole, 19, 21, 23, 26-27, 36, 213; Cave of the Vampires, 47, 48; Carlsbad Caverns, 153-54, *153;* Ney Cave, 213-17
  flying at lights, 212-13, 316; homing, 212; longevity, 223; maternity quarters, 288; migration, 58, 153-54, 223; numbers, 221; odor, 19, 247-50; protection, 211; rabies, 40, 57-58; superstitions, 207, 230; young, 47-50, 210; mentioned, 34, 37, 45, 106, 114, 121-22, 149, 152-53, *225,* 261, 264, 266. *See also* Echo-location; Free-tailed Bat; Guano; Hibernation; Vampire Bat; and names of other species

Battlefield Crystal Cave, Va., 131

Bear, 105, 174, 175, 185, 209, 271; bones, 113; Black, 177; Grizzly, 175, 177

Beaver, 173

Beaver Creek Cave, Tex., 209

Bedbug, related species, 211

Beebe, William, 236

Beetle(s), 211; Blind, 114

Bend, Ore., 17

Bibliography of North American Speleology, 1707-1950, Davies, 3

Big Horn Mts., 17

Bisbee, Ariz., 18

Bishop, Marshall, 257, 264-65, 266

Bishop, Stephen, 106, 108, 124

Bison (Buffalo) 105, 188, 190; Wood, 174, 176

Black Hills, S, D., 17

Black, Homer T., 142-57, *151*

Black Rock Cave, Conn., 298

Black Rock State Park, 298

Blake, John, 166

Blasting, 114, 160, 162-63, 176-77, 180, 182, 209, 228

Blatchley, W. S., 319

Blind Beetle, 114

Blind Salamander, 271-81, *274, 275* degeneration, 273; descriptions, 272, 275, 279; discovery, 271; eggs, 279-80; Europe, 271; eyes, 273, *274;* Georgia, 281; heredity, 276-77; larva, 273-74; metamorphosis, 273-74, *274, 275,* 276-77; Ozarks, 271-80; Texas, 280-81

Blindfish, in Mexico: adaptation, 256, 266; behavior, 258, 261, 266, 267; biology, 257, 266; cancers, 268; collecting, 53, 265; distribution, 267-68; eyes, 256-57, 264; food, 258-59, 261, 266. *See also Astyanax*

INDEX

in U.S.: degeneration of eye, 284; food, 285; Mammoth Cave, 108; Ozarks, 281, 282, 284; passage beneath Mississippi, 284; sensitivity, 284; size, 285

Bones: cleaning process, 175, 180; human, 148, 185-92, 193-205; stalagmitic covering, 189-91, 203-204. *See also* Archeological investigations

Bourglay, Jules. *See* Leather Man

Bower Cave, Calif., 98, 193

Brachiopod, 143

Bracken Bat Cave, Tex., 213, 216, 221

Breakdown, 13, 14, *15*, 163, 187, 189

Breathing cave, 13, 159, 160

Breder, Dr. Charles M., Jr., 53, 257-62, 264-67

Bridges, William, 256-68; co-author, *Snake-Hunter's Holiday*, 256

British Columbia, 18

British Guiana, 234

Brown, William, *97*, 97-102, 103

Brucker, Roger W., 78-89, 158-71

Bryan, Kirk, 191, 204

Bryozoan, 143

Burdon, Robert, 166, 167

Burgess, Major Richard, 151

Burlington, Conn., 299

Burson, Sen. Holm O., 150

Bushey's Cavern, Md., 183

Calaveras County, Calif., 194-97, 201

Calcite: rate of deposition, 203-204; source, 18; mentioned, 9, 13, 34, 131, 134. *See also* Mineral deposits

Calcium carbonate, 4, 9, *10*, 143, 189, *190*, 319, 323

Calcium cups, 263-65, *263*

Calcium nitrate, 13, 106

Calcium sulphate, 4. *See also* Gypsum

California, 4, 5, 17, 90-91, 95, 98, 102, 193-204, *196, 199*

Cambarus ayersii, 282

Camel, American, 91, 188

Cameron, Archie K., 313

Campbell, Charles A. R., 211

Canada, 18

Canyon de Chelly, Ariz., 5

Cape Maisí, Cuba, 243-55

Caracas, Venezuela, 234

Carbon-14, 185, 192, 203-204

Caripe, Venezuela, 234

Carley, Addison, 199, 203

Carlisle, Pa., 228, 232

Carnegie Cave, Pa., 232

Carlsbad, N. M., 149, 221

Carlsbad Caverns, N. M., 17-18, 142-57, *146, 147, 150, 151, 153 155, 156,* 208, 211, 213, 216-17. *See also* Development, Origin, National Parks

Carvings on formations, 137, 252-53, *252, 253*

Cataract Gulch Cave, Calif., 195

Cave(s): areas, *x,* 16-18, 130-31 (*See also* under names of states); depth, 18, 24, 86, 100, 142, 148; gypsum, 4, 17-18; lava, 4, 5, 17; talus, 5; tunnel, 137-40, 187

uses: barracks, 130, 132, *137;* counterfeiting, 321; entertainment, 306-307; habitation, 185-92; liquor vault, 306; outlaws, 304-12; worship, 235, 253.

See also Commercial; Development; Fauna; Formations, Origin

Cave City, Calif., 196

Cave Cricket, 114, 285, 288, *288*

Cave in Rock, Ill., 304-12, *305*

Cave of Skulls, Calif., 195, 198-9, *199*, 201

Cave of the Catacombs, Calif., 195-198, *196*

Cave of the Lost Maiden (Samwel Cave) Calif., 198

Cave of the Vampires, 39-58, *43, 46*

Cave of the Winding Stair, Calif., 91-92

Cave rat, 114, 175, 191

Caverns of West Virginia, 130

Cave Salamander, 280

328   CELEBRATED AMERICAN CAVES

Caves Beyond, The, 78
Cavetown, Md., 183
Cave-without-a-Name, Tex., *10*
Centipede, 248
Ceuthophilus, 288
Chambersburg Valley, 130
Chamberlain, Fred, 299
Chamberlain, Walter S., 95
Characin. *See* Blindfish, Mexico
Charcoal, 183, 185, 192, 318, 322
Charlestown, W. Va., 132
Charlet, M. L., 109, 114
Charmichael, Henry, 168-69
Chehalumche, troglodytic giant, 195, 197-98
Chesapeake and Ohio Railroad, 140-41
China, 185
Chow Kow Tien caves, 185
Circulation: air, 13, 101, 103, 249, 279, 293; water, 5-8, 144
Clarksburg, W. Va., 136
Claustrophobia, 45
Cleaves, Arthur B., 226, 228
Clemens, Samuel, 313-16
Cleversburg Sink, Pa., 232
Cliff dwellings, 4, 5
Climate, 173, 175, 188-92, 208
Cluster Bat, *217*
Cockroach Cave. *See* Cueva del Agua.
Cockroaches, 248, 250-1, 253
Collapse, 13-14, 31-32, 145, 176, 179, 183, 228. *See also* Breakdown
Collecting, 15, 41-42, 53, 56, 263, 265-267, 282
Collett, Prof. John, 318
Collins family: Floyd, 157-71, *159, 164, 171;* Homer, 165-66; Lee, 165; Marshall, 165
Colombia, 56
Colorado, 5
Colossal Cave, Ky., 160
Column, *10, 137,* 242

Commercial caves, 130-31, 160, 269 *See (Calif.)* Mercer, Moaning; *(Ky.)* Floyd Collins Crystal, Mammoth, Mammoth Onyx; *(Ind.)* Wyandotte; *(Mo.)* Mark Twain, Marvel; *(N. H.)* Polar; *(N. M.)* Carlsbad; *(Ohio)* Ohio Caverns; *(Tex.)* Cave-without-a-Name; *(Va.)* Battlefield Crystal, Dixie, Endless, Grand, Luray, Massanutten, Melrose, Natural Tunnel, Shenandoah, Skyline; *(W. Va.)* Organ; no longer operated:*See* Colossal, Fountain, Madison, Salts, Weyer's
Communication: 36, 83, 182; signals, 61, 64-67; underwater, 103
Confederate Army, 136
Connecticut, 290-303
Conodoguinet Cave, Pa., 228-29, 232-233
Conservation, 15, 95-96, *96,* 211, 225
Cope, Edward D., 184, 272
Copper, 18
Coronado, Sr., 262, 265, 266
Corrigansville, Md., 173
Couffer, Jack C., 213, 216
Cox, J. L., 162
Craighead Cave, Pa., 232
Crane, H. R., 192
Craters of the Moon Nat. Mon., 5
Crayfish: Cueva Chica, 262, 264; white, 281-2 *282,* 285, 287
Cretaceous, 144
Crocodile, 180
Croghan, Dr. John, 116-19
Crustaceans, 33, 285
Crystal Cave, Ky. *See* Floyd Collins Crystal Cave
Cuba, 18, 243-55
Cueva Chica, Mexico, 256-68, *260, 263, 265, 267*
Cueva del Abra, Mexico, 57
Cueva del Agua, Cuba, 243-55, *246, 252, 253*
Cueva del Carrizal, Mexico, 57
Cueva de los Sabinos, Mexico, 39-58, 267

INDEX 329

Cueva del Pachon, Mexico, 57, 267
Cueva la Boca, Mexico, 57
Culverwell, Tom, 59, 65
Cumberland Bone Cave, Md., 172-84, *177, 181*
Cumberland, Md., 172, 175
Cyprinodon diabolis, 94-95, *97*

Dalquest, Walter, 56, 58
Danehy, Edward, 193-99
Dating techniques, 192, 203. *See also* Carbon-14
Davidson, Charlie, 278-79
Davies, William E., 3-18, 130-41
Davis, Kenneth, 186
Davis, Dr. W. M., 144
Dearolf, Kenneth, 114, 281-83
Death Valley, Calif., 91-95
Death Valley Nat. Mon., 96
Deep Hollow Cave, Conn., 291
Deer, 173
Delaware, 16; River, 183
Derriengue, 57-58
Desmodus rotundus, Vampire Bat, 39-58
Development, 6-8, 130-31, 143-46 anastomoses, *6;* below water table, 6; drainage, 6-7, 145; pattern, 6, 8, 16-18, 141, 144; stages, 6-7, 144-46; subsurface erosion, 5; theories, 5-8, 144. *See also* Formations
Devil's Hole, Nev., 90-104, *96, 97, 99*
Devil's Hole Cave, Nev., 96, *97,* 100-104, *100, 101, 103, 104*
Devil's Sinkhole, Tex., 19-38, *20, 27, 31,* 91, 212-13, 216, 218
Devonian, 173
Diablotin (Guacharo), 238
Ditmars, Raymond, 256
Diurnal rhythm. *See* Experiments
Diving: Devil's Hole, 90, 95-104, *99;* equipment, 95 - 104, *106;* hazard, "bends", 98-100, 102, 104
Dixie Caverns, Va., 131

Dome pit, 8, 17, 86
Dominica, 13
Doyle, Bee, 162-65
Doyle, Jewell, 164-65
Drake, Dan, 118-19
Dripstone, *8,* 34, 93, 112
Dug Way Cave, Conn., 291, 297
Dunton, Samuel C., 257, 266
Durham Cave, Pa., 183
Dyer, James, 82, 87-89

Early Man, 5, 91, 185-92, 204-205 China, 185; Europe, 185, 191; pottery, 187, 189; tools, 187-88, 191; weapons, 186, 188-92. *See also* Indians; Footprint
East Berlin, Conn., 299
East Haven, Conn., 300
Echo-location: experiments with bats, 224; Guacharos, 239; failure of, 52, 213, 264
Echo River, Mammoth Cave, 12, 106-7, 114, 124
Ecuador, 234
Ecubut Hill Cave, Conn., 291
Edwards Plateau, Tex., 17, 56, 209
Eddy N. M. *See* Carlsbad
Ehman, Burnett, 82, 85
Eigenmann, Carl, 272-4, 284; *Cave Vertebrates of North America,* 273
El Candela, Mex., 57
Elephant, American, 173, 192
Elevator, 149, 154, 156-57
Elk, 105, 174, 180
Endless Caverns, Va., 131
Enlargement, 13-14, *15,* 144-45. *See also,* Breakdown, Collapse
Entrance scenes, *20, 25, 31, 43, 60,* 62-3, map, *96, 97, 99, 138, 153, 156, 162, 174, 177, 219, 225, 227, 235, 260, 277, 305*
Equipment: archeology, 189; banding, 230-31; boating, 109, *110,* 111, 262; Bone Cave, 179-80; bosun's chair, 24-

26, 28-30; Cave of the Vampires, 43; Cueva Chica, 257; diving, 95-104, *96;* idol removal, 254-55; rescue, 165-68; rigging, *20,* 21-38, *27;* rock-climbing, 60-77, 79-89. *See also* Accidents.

Estes, Edward, 162, 164-65

Eumops, 218

Europe, 185, 270-71

Eurycea lucifuga, 280

Evaporation, 9, 144

Experiments: bats, 239; fish, 266-68; darkness, 239; diurnal rhythm, 119-21; egg-laying, 279-80; Guacharo, 239; metamorphosis, 276-77; tuberculosis treatment, 116-19

Explorers: French, 269, 304; Spanish, 148, 269

Explosion, 43, 209, 221

Falling Waters, W. Va., 180, 182

Fatigue, 74, 75, 76, 77, *84,* 86, 102, 160, 165, 167, 170

Fault, *7,* 145

Faun Rock, Conn., 291

Fauna, 274-89
  adaptations, 256, 266, 268, 285, 288; aquatic, 174, 285; distribution, 284; environment, *274,* 276-77, 287-89; evolution, 289; food, 286, *288,* 288-89; research, 289. *See also* Bats, Blind Salamander; Blindfish; Crayfish; and names of other species.

Fertilizer (guano), 208-11

Fill, 7, 13, *14,* 18, 133, 148, 175-76

Fire, 208-209, 221

Fish, Desert Spring (pupfish), 94, 95, 97. *See also* Blindfish

Fishes, Cueva Chica series, 256, 265-66, *265;* behavior, 267; evolution, 259; interbreeding, 259, 266. *See also* *Astyanax,* Blindfish

Flare-throwing. *See* Illumination

Flatworm, 103, 286, *287*

Flies, 27, 211, 288

Flint Ridge, Ky., 161

Florida, 17

Flowstone, 34, 88, 93, 112, 117, 203-204, 323

Floyd Collins. *See* Collins family

Floyd Collins Crystal Cave, Ky., 160-62, *171*

Fluger, Colonel, 307-08

Fluorite, 13

Folklore, 17

Folsom, N. M., 188

Folsom: culture, 190, 191; "points", 188

Folsom Man, 185, 188, 191

Foote, Leroy W., 290-303

Footprint(s), human, 113, *126,* 136, 229, 317, 323

Ford's Ferry band, 312

Forestville, Conn., 294

Formations: *11,* 13, 142-43, 146, *146, 147, 151,* 152, *155,* 240-41, 320-23; age of, 144, 321, 323; color of, 34, 142-43; deposit on, 147, *147;* origin, 9, 10; varieties, 146. *See also* Anthodite, Aragonite, Bacon rind, Calcite, Column, Dripstone, Flowstone, Gypsum, Helictite, Iron carbonate, Rimstone, Shield, Stalactite, Stalagmite, Travertine

"Fossil" cave, 18

Fossils, 13, *14,* 91, 133, 175-84, *174, 180,* 185-92, 270-71. *See also* Mummified remains

Fountain Cave, Va., 132

Fox, 174, 271

Franklin, W. Va., 3, 5, 7

Frankstown, Pa., 183

Free-tailed Bat, 27, 152, *153, 207,* 211, 215, 219, 220; former abundance, 206-208

Fremont, General, 136-37, *137*

Friedman, Ralph, 257, 261, 266

Frio Cave, Tex., 209-10, 213, 216

Frog, 174

Front Royal, Va., 133

Frozen Niagara, 107, 112

Fruit Bat, 41, *48,* 51, 58

INDEX 331

Fungi, 4
Fungus gnats, 285

Galambos, Robert, 224, 239
Gamaliel's Den, Conn., 291
Garvin, William, 124
Gas: ammonia, 211; carbon dioxide, 9; volcanic, 101
Gauche, Juan, 249, 251, 254
General Land Office, 149
Georgia, 135, 281
Gerald, John, 166-68, 170
Germany Valley, W. Va., 141
Gidley, J. W., 176
Giron, Dr. Alfredo Tellez, 54, 57
Glaciation, 18, 147, 173, 175, 181, 184, 188-89, 191; effect on caves, 18
Golondrinas Mines, Mexico, 57
Gordon, Dr. Myron, 257, 258, 259, 261
Grand Caverns, Va., 131, *132*
Grasshopper. *See* Cave Cricket
Gray Bat, *285*
Great Basin, 90
Great Crystal Cave, Ky. *See* Floyd Collins Crystal Cave
Great Saltpeter Cave, Ky., 139
Greenbriar River, W. Va., 140
Greenville Saltpeter Cave, Ky., 136
Green River, Ky., 105-07, 115, 206, 310
Greenwich, Conn., 243, *302*
Gresser, Dr. Edward Bellamy, 257, 258, 265, 266, 267
Griffin, Dr. Donald R., 223-24, 236, 239
Grottoes, Va., 131
Ground sloth, 91, 133, 148, 185, 187, 190-91, 203, 245, 247
Ground water, 5, 142, 144-45
Guacharo, 234-42, *235, 237, 240, 241,* 256
captivity, 238; conservation, 234, 236; description, 236-37; distribution, 234, 238; echo-location, 239; eyes, 237; flight, 235, 238-39, 242; food, 238;
guano, 238, *241;* nesting, 235-36; numbers, 234; odor, 237; slaughter (oil-harvest), 235-36, *235,* 242; voice, 235, 238-42
Guacharo Cave, Venezuela, 234-42, *235, 237, 240, 241*
Guadeloupe Mountains, 142, 145, 148-49
Guano: bat, 36-37, 50-51, 250; mentioned, 33, 45, 189, 208; description, 210-211; explosive nature, 208-209; fertilizer, 208; food for aquatic life, 53, 285-86; Guacharo, 238, *241;* mining, 24, 106; odor, 211. *See also* Gunpowder; Saltpeter
Guano Bat. *See* Free-tailed Bat
Guatemala, 18
Guides, 105-16, *107, 109, 110,* 123-125, *123, See also* Naturalists, Park
Gunpowder, 106, 121, 123, 208-09, 211
Gypsum, 4, 13, 17, 18, *112, 113,* 126-129, 144, 147, 160, 163
Gypsum Cave, Nev., 91, 243

Hagerstown Valley, 130
Halliday, William R., 90-104, 193-205
Hannibal, Mo., 313-16
Hanks, Henry G., 194-5, 198
Hansen, Carl, 108-10, *110,* 113
Hansen, Pete, *109,* 111, 113
Harpe, "Big" Micajah, 308-12
Harpe, "Little" Wiley, 308-12
Harper's Ferry, W. Va., 176
Harrington, M. R., 243-55, 301, *302*
Harrisonburg, Va., 136
Hartford *Courant,* 299
Harwinton, Conn., 292
Hawkins, Annie Laurie, 315-16
Hawks: Coopers, 215; Duck (Peregrine Falcon), 214, 216; Kestrel (Sparrow), 215; Red-tailed, 215; Sharp-shinned, 215; Swallow-tailed Kite, 214
Haworth, George, 179
Hay, R. D., *263*
Haynes Cave, W. Va., 135

Hedricks Cave, W. Va., 133-34
Helictite, 10, *11*, 34, 160
Hellhole, W. Va., 141
Hermit. *See* Leather Man
Herrera, Miguel (Mike), 39, 42, 44, 45, 46, 52
Herrera, Malaquinas, 42, 44, 46
Hershey Farm Caves, Pa., 232
Hibben, Dr. Frank C., 185-92
Hibernation: bats, 224-33, *225, 229, 231, 285,* 288; conditions required, 226
Hittell, John S., 194
Hoffmaster, Richard, 182
Holley, Robert A., 149-50
Holston: River, 213; Valley, 131
Hoppin, Ruth, 281, 284
Horror Cave. See Cueva del Agua
Horse, 188
Hoskins, R. Taylor, 114
Hotchkiss, Chauncey, 294
Houchins, 106
Hovey, E. O., 272-73
Hovey, Horace C., 318
Hubbs, Dr. Carl L., 94-95, 284
Hubbs and Innes, description of Blindfish, 258
Huckleberry Finn, 313
Huckleberry Hill Cave, Conn., 291
Hudson River, 291, 303
Hulstrunk, William, 82, 85
Humboldt, Alexander von, 41, 236, 239; National Park, 236
Humidity, 13, 103-04, 124, 133, *274*, 289
Hungerford, Celester, 299
Hunt, Claude, 111, 113
Hunt, Leo, 108-10, *109, 110,* 113
Hurricane Island rapids, 306-07
Hutchings, J. M., 194
Hyde, Roy, 168

Ice Age. *See* Glaciation
Ice caves, 5
Idaho, 5, 17
Idol of Cape Maisí, 243, *253, 255*
Illinois, 18, 304-12
Illumination: acetylene lanterns, 187; candles, 44-45, 53, 316; carbide, 44, 52, 60; electricity, 154-56; electric miner's lamps, 44; flare-throwing, 123-24, *123;* flashlight, 52, 257; gasoline lantern, 44, 52, 80, 108-109, 212-13; hazards, 52, 64, 86, 161; kerosene lanterns, 152, 154, 158, 160-61, 248; pine fagots, 134-35; red fire, 272; torch, 129, 187, 201, 270, 321; underwater, 103
Indian Cave, Tenn., 213
Indian(s): Apache, 148; Basket-maker, 148; Chaima, 234-42; Cherokee, 317; Delaware, 317; Ciboney, 244, 246, 253; excavation of stalagmite, 317-23, *319;* footprint, *126,* 323; gypsum miner, 124-29; legends, 197-98; Lenape, 183; Mexico, 40-53; Miwuk, 195, 197; mummy, 124-29; *127, 128,* 198; Shawnee, 317; Taino, 244-45, 253. *See also* Artifacts; Bones
Indiana, 17, 317-23, *319, 322*
Indiana Geological Survey, 318
Indiana University, 273
Industry, 4
Inger, Robert, 284
"Injun Joe", 315
Innocents Abroad, Mark Twain, 313
Insect(s), 41, 53, 148, 208, 211, 228, 286, *288*
Iron carbonate, 227
Iron oxide, 142
Isopod, 27, 278, 285, *286,* 287

Jackson, George F., 304-12, 317-23
Jackson, Stonewall, 130, 137
Jamaica, 18
James Cave, Ky., 78-89, *80, 84, 87, 88*
James, Frank, 78-79, 89
James, Jesse, 78-79, 89

INDEX

Jefferson, Thomas, 130-34
Jericho Rock, Conn., 291, 298
Jewell Cave, W. Va., 140
Johnson, F. M., 27
Johnson, Harald N., 57
Joints, 16, 132, 144, 148
Jones, David, 85
Jones Quarry Cave, W. Va., 181
Juniper Cave, Calif., 193

Karst, *16*, 17
Kay, Le Roy, 180
Kelly Farm Caves, Pa., 232
Kentucky, 12, 17, 78-89, 105-15, 116-29, 139, 158-71, 206, 284
Kezer, James, 279
Kittatinny Tunnel, Pa., *225*, 226-33
Kleitman, Dr. Nathaniel, 119-21

Laboratory "cave", in museum, 266-67
Lake Saltonstall, Conn., 300
Laron, Marguerite, 290-91
La Patana, Cuba, 243-55, *246, 252, 253*
Las Huertas Canyon, N. M., *186*, 187
Las Vegas, N. M., 90, 91
Laurel Creek Cave, W. Va., 6, 9, *15*
Lawhorn, Richard, 104
Lava Beds Nat. Mon., Calif., 4, 5
Lava caves, 4, 5, 17
Lava River State Park, Ore., 5
Lead, 13, 18, 270
Leather Man, 291-302, *293, 302*
Leatherman's Cave, Conn., 243, *302*
L'Echo du Pacifique, 200-201
Lee, Willis T., 142, 151, 152
Leiba, Gaspar, 247, 248, 249, 251, 254
Leinhaupel, Frank, 104
Life, 216
Lilburn Cave, Calif., 102
Limestone, 4-18, *16*, 28, 131, 142-48, 161; lead-bearing, 270; pits in, 141; mentioned, *84,* 92-93, 129, 137, 172, 173, 190, 210, 269, 304
Lindbergh, Charles A., 170
Lindbergh, Jon, 98
Little Brown Bat, 27, 106, *229, 231*
Lix, Henry M., 105-15, 124
Long-eared (Lump-nosed) Bat, 27
Long, "Bige", 149
Long-tailed Salamander, 276, 279
Lorenz, Robert, 104
Los Angeles, Calif., 243
Los Sabinos, Mexico, 39, 42, *43*
Lost John, 127-29, *127, 128*
Loud, L. L., 243
Louisville, Ky., 77, 80, 116, 165, 166, 307; Fire Department, 166
Louisville *Courier-Journal,* 166
Louisville *Herald,* 166
Lovelock Cave, 243
Lower Salt River Cave, Tenn., *11*
Luray Caverns, Va., 131
Lynch sisters, 277
Lynx, 114, 271

McCloud River area, Calif., 198
McClure, Roger, 81-82, 88, 89
McDowell Cave., Mo., (Sims, Saltpeter and Mark Twain Cave): bat colony, 316; buried treasure, 315; cadaver experiment, 314; description by Clemens, 316; storage of cannon, 314
Madison Cave, Va., 132
Mammoth, 185, 188-89, 191; tusk, 192
Mammoth Cave Ridge, Ky., 161
Mammoth Cave, Ky., 12, 105-15, *107, 109, 110, 112, 113,* 116-29, *117, 122, 123, 126, 127, 128,* 160, 162, 206, 225, 279, 310
Mammoth Onyx Cave, Ky., 289
Man. *See* Early Man, Folsom Man, Leather Man, Pekin Man, Sandia Man
Manganese dioxide, 13
Mannix, Dan, 39-58

Mannix, Jule, 39-58
Maps, x, *62, 63,* 82, 88, *140, 186, 190, 314*
Marble Cave. *See* Marvel Cave
Mark Twain (Samuel Clemens), 313-16
Mark Twain Cave, Mo., 313-16; map, 314
Marten, 114
Martinsburg, W. Va., 180
Marvel Cave, Mo., 270-81, *277, 285, 286*
Maryland, 130, 172-84
Mason, Samuel, 305-08, 312
Massanutten Caverns, Va., 131
Mastiff Bat, 218
Mastodon, 173, 175, 180, 185, 188, 191
McDowell, Dr. E. D., 314
Megalocnus. See Ground Sloth
Melrose Caverns, Va., 131, 136, 137, *137*
Mercer Caverns, Calif., 193, 195, 203
Mercer's Indian Burial Pit, Calif., 195
Merriam, C. Hart, 195
Metamorphosis. *See* Blind Salamander
Mexican Dep't. of Fisheries, 257, 259
Mexico, 18, 39-58, 256-68
Mice, 174; deer, 175
Midden, 148, 183, 245
Middletown, Conn., 299
Mildrum, E. W., 299
Miller, Jerry, 195
Miller, Robert R., 94
Miller's Cave (Cave of the Catacombs), 195-97
Miller, William Burke, 166-70
Mineral deposits; 203-204; 269-70. *See also* Travertine
Mineral(s), 13, 18, 143-44, 148. *See also* Aragonite, Calcite, and names of other minerals.
Miners: prehistoric, 125-29, *126*, 318-23; for onyx, 79. *See also* Saltpeter, miners

Mines, 5, 14, 218, 223
Mink, 174
Missouri, 17-18, 79, 269-89
Mite, 211
Mt. Higby, Conn., 299
Moaning Cave, Calif., 193, 195, 198-201, *199,* 203-204
Mohr, Charles E., 39-58, 206-22, 223-33, *233,* 269-89, 301
Mojave Desert, Calif., 91, 95
Mold, 4, 116
Monterrey, Mex., 56
Moose, 174-75
Morrow, Rep. John, 150
Mosqueras, Cecilio, 246, 247-48, 249
Moth, 288
Mother Lode Country, 193-205, *199*
Mountaineering, 90. *See also* Rock-climbing
Mueller, Peggy, *134*
Mule-eared (Pallid) Bat, 219
Mummified remains, 271, 273. *See also* Indian, mummy
Murcielegos (bats), 39
Murders. *See* Cave in Rock
Murrina, 57
Museum of the American Indian, Heye Foundation, 244-55
Muskrat, 173
Myotis grisescens, 285
M. l. lucifugus, 229, 231
M. sodalis, 217
Mysz, Fred, 82, 85
Mythical caves, Cyclopean, Colo., 198; Magnetic, Calif., 198

Naica, Mexico, 18
National Geographic Magazine, 152
National Geographic Society, 152
National Parks and Monuments: Carlsbad Caverns Nat. Park, 142-57; Craters of the Moon Nat. Mon., 5; Death Valley Nat. Mon., 95-96; Lava

Beds Nat. Mon., 4, 5; Mammoth Cave Nat. Park, 105-15, 116-29
National Speleological Society
Chapters: Central Ohio, 82, 89; Denver, 90; Kentucky-Indiana, 89; Pittsburgh, 181-82; Salt Lake City, 90; Seattle, 90; Southern California, 95, *97*, 99, 103, 193; Stanford, 193, 197
protection of caves by, 97; mentioned, 3, 116, 172, 195, 199, *263*, 313, 317
Natural History, 125, 129, 277
Naturalist(s), Park, 105, 142, 153-54
Natural Tunnel and Chasm, Va., 131
Naugatuck River, Conn., 298
Neely, Peter M., 91, 95
Nevada, 90-104, 96, 97, 99, 185
New Brunswick, 18
New Cave, N. M., 142, *150*, 208
New Discovery, Mammoth Cave, *112*, 114-15, 124
New Hampshire, 4
New Haven *Register,* 300
New Jersey, 16
New Mexico, 5, 16, 142-57, 185-92, 204, 208, 211, 213, 216-17, 219
New Orleans, La., 106
New York *Daily News,* 292
New York *Sun,* 256
New York Zoological Society, expedition, 256-68
New Zealand, 3
Ney Cave, Tex., 210, *210*, 213-14, 216, *219*, 221
Nicholas, Brother G., 72-84
Nitrate, 106, 121-23, 208-209
Noble, G. Kingsley, 274-77
Norfolk and Western Railway, 131
Norristown, Pa., 173
Nova Scotia, 18
Number of caves, *x*, 4, 16

Oakville, Conn., 298

Ochre, 189-90
Odor: bats, 19, 247-50; Guacharos, 237; guano, 211; vampire pools, 49-50
Ohio, 130, 136, 137
Ohio Caverns, Ohio, *11*, *12*
Ohio River, 304-12, *305*, 317
Oklahoma, 17, 269, 285-87
Olm, European Blind Cave Salamander, 271
O'Neill Cave, Calif., 195
Ontario, 18
Onyx, 79, 107, 270
Operation X-ray, 216-22
Opossum, 271
Oregon, 5, 17
Organ Cave, W. Va., 133-34, *134*
Origin, 5-8, 130, 131, 142-48. *See also* Development; Enlargement
Orr, Phil C., 203-204
Osceola, W. Va., 138
Ossining, N. Y., 303
Otter, 173, 271
Outcropping. *See* Karst
Outlaws, 78, 89
Owl: Barn, 215; Great Horned, 215; Screech, 278
Ozark Red-backed Salamander, 289
Ozark Mountains, 269-89

Pack rat. *See* Cave rat
Paleolithic Age (Stone Age), Europe, 191
Paleontological investigations, 172-84, 198
Palmer, Robert, 82
Pan-American Highway, 39
Panther, 271
Parasites, 211. *See also,* Mite
Passenger Pigeons, 206-208
Peccary, 133, 180, *181*
Pecos: River, 148; Valley, 149
Pekin Man, 185

Pennsylvania, 7, 130, 175, 183, 216, 223-33, 279
Pennsylvania Turnpike Tunnels, 226-233, *225, 227*
Permission to enter, 15
Permian, 143-44
Peru, 234
Petroglyph, 136, 252, *252*
Phelps, William H., 236
Phelps, William H. Jr., 236
Photography: Collins' body, 170; Idol, 253,54; special tourist trip, Carlsbad Caverns, 157
Phreatic cycle, 144, 147. *See also* Development
Pictograph, 136, 148
Pietri, Eugenio de Bellard, 234, 236, *237*, 242
Pillar of the Constitution, 318-22, *319*
Pipe bridge, 81-6, *87, 88*
Pipe line, 122, *122*, 123
Planarian. *See* Flatworm
Plantlife, Pleistocene: birches, 173; duckweed, 174; maples, 173; water lilies, 174
Pleasantville, N. Y., 301
Pleistocene, 95, 145, 147, 176-84, 189
Pocket Gopher, 180
Polar Caves, N. H., 4
Pond, W. Alonzo, 125-29
Porcupine, 173
Porte Crayon, 138-39
Port Kennedy Bone Cave, Pa., 183
Potassium nitrate, 121-23, 134
Potomac Valley, 174
Potter, Floyd, *20*, 21-38
Powell, Truman, 270, 272
"Pterodactylus", 271
Predators, 213-14, 226, 233
Puerto Rico, 18
Pujal, 259
Pulitzer Prize, 170

Puma, 173, 180

Quarrying: calcium carbonate, 318-20, 322; flint, 318, 321-23

Raccoon, 114, 213
Radar. *See* Echo-location
Radiocarbon. *See* Carbon-14
Raney, Ed, *20*, 21-38
Rasquin, Priscilla, 268
Rattlesnakes, 22-23
Research: bibliographic, 194-95; opportunities, 154, 289, 323. *See also* Experiments
Reville, Dorothy, *217*
Rey, Don Antonio, 245
Richards, Tom, 3
Richardson, Bruce, 119-21
Riegelsville, Pa., 183
Rigging, *20*, 21-38, *27*
Rio Grande, 57
Rio Tampaon, 258, 264, 266
River Pirates, 304-305
Roaring River, Mammoth Cave, 106, 108, *110*, 114
Roark Mountain, 270
Rock: engineering properties, 4; insulation, 5; stability, 14. *See also* Breakdown; Collapse; Limestone
Rock-climbing, 19, 59-77, *60, 67, 68, 70, 71, 72, 73, 75*, 80-89, *81, 84, 85*, 141; signals, 61, 64-67
Rock House Cave, Mo., 271
Rocksprings, Tex., 22-23, 26, 212
Rocky Mountains, 17, 90, 144, 188
Rodriguez, Dr. Victor J., 245, 247, 249, 250, 251
Rogers, Joseph F., 311
Rogers, Nancy, *6*
Rohl, Dr. Eduardo, 236
Ronceverte, Va., 133, 140
Rose Cave, Tex., 218
Rosenbloom, Libby, 268

INDEX 337

Rothrock, Charles J., 323
Ruffed Grouse, 180
Ruhoff, Theodore, 177
Ruthven, A. G., 56

Sabre-Tooth, 91
Sacramento *Union*, 196
Safety: diving, 97-104; helmet, 182; rock-climbing, 59-77, 80-89, *84*, rules, 15. *See also* Accident, Equipment, Illumination, hazards
St. Lawrence Valley, 18
Saltpeter mining: caves, 106, 135, 209-210, 270; cost 121; health of miners, 116; leaching, 122-23; quantities, 121, 209-10; "saltpeter monkey", 134; relics *122, 134,* 135-36, *135;* vats (hoppers), 122, *122,* 134, *134*
Samwel Cave (Cave of the Lost Maiden), Calif., 198
San Antonio, Tex., 22, 56, 209-10, 215
Sand Cave, Ky., 162-171, *159, 162, 164, 169*
Sanderson, Ivan T., on Guacharo, 237-39
Sandia Cave, New Mex., 185-92, *186, 190,* 204
Sandia Mountains, 186-87
San Francisco, Calif., 197-98
San Jacinto Cavern, Mexico, 56
San Luis Potosi, 56, 256-61
Santa Barbara Museum of Natural History, 203
Sarcoxie, Mo., 271, 281
Saskatchewan, 188
Saw Mill River Cave, N. Y., 292
Sawtelle, Ida V., 4, 59-77, 141
Schoolcraft, Henry, 270
Schoolhouse Cave, W. Va., *14,* 59-77, 60, 62-63 (maps), *67, 68, 70, 71, 72, 73, 74, 75, 78,* 141
Scientific American, 271, 272
Scorpion, 248; whip, 263
Sea cave, 4
Sediment, 4, 18, 143-45. *See also* Fill
Selenite, 18. *See also* Gypsum

Shelters, 185-92, 270. *See also* Leather Man
Shenandoah Caverns, Va., 131
Shenandoah Valley, 121, 131
Shepherd of the Hills, Wright, 271, 278
Sheridan, General, 137
Shields, 131, *132*
Shippensburg, Pa., 232
Shouse, Henry C., 312
Shrew, 174, 179
Shrub Oak, N. Y., 292
Sickness, 266
Sierra Nevada Mts., 90, 193-205, *199*
Signatures, 86, 89, 136-37, *137,* 196-97
Silica, 180
Silt. *See* Fill
Simmons, Edward, 97, 102
Sinaloa, Mexico, 58
Sinkhole, 19-38, *20, 27, 31,* 85, 160, *174,* 174-76, 182, 194-95, 200, 265, 272
Sinks of Gandy Creek, Va., 137-38, *138,* map 140
Skeleton(s), 174, 185, 192-93, *196*
Skull(s), 124, 182, 185, 193-205; with gold tooth, 195
Skunk(s), 174, 213
Skyline Caverns, Va., 131, 133
Slaughter Canyon, N. M., 142
Slimy Salamander, 289
Sloane, Howard N., 116-19, *217, 233,* 234-42, 313-16
Smith, Philip, 82, *84,* 85, 86, 87
Smithsonian Institution, 271, 273
Snake, 180, 251
Soda Straw. *See* Stalactite
Solomon's Hole (Ossiferous Cavern), Calif., 200
Solution, 4, *6,* 144, 145
Sonar. *See* Echo-location
Sorrel, Farmer Clem, 292
Sotano de la Arroya, Mexico, 267
Sotano de la Tinaja, Mexico, 267

Sounds magnified: bats in flight, 212, 248, 250-51, 264; bird cries, 43; flashbulbs bursting, 46-47; Guacharo cries, 235, 238-42; gunfire, 53; waterfall, 12, 283; waves lapping, 108

South Penn Railroad, 226

Sparta Cemetery, Ossining, N.Y., 303

Speleology. *See* Research

Speleothem (formation), 203-204

Sphalerite. *See* Zinc

Spider(s), 243, 263, pholcid, 260

Spitler, David, 85

Splerpes (Eurycea) melanopleurus, 271

Sponge, 143

Springfield, Mo., 270, 281

Sprunt, Alexander, Jr., 214-16

Squirrel, 173

Stager, Kenneth, E., 214-15

Stalactite(s): carvings on, 137; origin, 9; mentioned, *12,* 23, 34, 146, 232, 240, 242, 316

Stalagmite(s): *10,* 146, *146,* 147, *147,* 252-54, *253;* Pillar of the Constitution, 318-20, *319;* Rock of Ages, 157; mentioned, *12,* 88, 242, 246, 316. *See also* Dating techniques; Idol

Stanislaus River, 199, *199*

Staunton, Va., 131

Steatornis caripensis, 234-42, *236*

Stejneger, Leonhard, 271

Stone, M. E., 26

Stone, Dr. Ralph W., 226-27

Stony Creek, Va., 140

Stratigraphy, 188-91, *190*

Sublett, Rolth, 149

Superstitions, 235-36, 242; bats, 207, 230

Surveying, 82, 85, 88, 151-52, 175, 218

Taconic Hills, N. Y., 291

Tadarida mexicana, 27, 58, 206, *207*

Tamaulipas, 267

Tapir, 175, 180

Tauste, Father, 234, 236

Taxco, Mexico, 39, 42

Temperature: bat flight, 26; constant, 133, 279; fluctuation in humans, 120-21; frigid, 208; frozen bats, 219-20; hibernation, 226, 228; hot spring, 90, 93, 98, 103, 104; range in U.S., 13; salamander distribution, 276, 288-89

Tennessee, 17, 130, 135 213

Texas, 17, 18, 19-38, 56, 91, 135, 188, 206-22

Texas Bat-cave Owners' Association, 211

Thatcher, Becky, 315

Thompson, Miles, Jr., 182

Thoms, W. P., 293

Thousand Rooms Cave, Ky., (James Cave), 79

Tories Den, Conn., 291, 299

Tourist parties, *107,* 152, 154, *155, 156, 157.* *See also* Commercial caves

Tracks. *See* Footprint

Trap, for fauna, *174,* 175, 194-95, 200, *See also* Sinkhole

Trask, John B., 194, 200-202

Travertine: bones beneath, 208; dam, *107,* 112; layers, 203-204

Treasure, 312, 315

Trinidad, 56-57, 256

Troglichthys rosae, 284

Tuberculosis treatment: 116-19; cottages, 117-18, *117*

Tunnels. *See* Pennsylvania Turnpike

Turkey Mountain Cave, Conn., 291

Turtle, 180

Typhlichthys subterraneus, 108

Typhlomolge, 281

Typhlotriton spelaeus, 272-80, *274, 275*

Ulmer Fred, 228-29, 232

Vadose cycle, 145. *See also* Development

Vallecito, Calif., 200-201

Valles, Mexico, 39, 42, 256-68

Vampire Bat: 39-58, *46, 50, 52* attacks, 40, 56-58; Dalquest on, 56,

58; distribution, 39, 56-57; excrement, 49-52; feeding, 53-56; flight, 49, 53; rabid, 52, 57-58; teeth, 51, *52*, 55; young, 49-50, *50*

Vampire legend, Slavonic, 39

Vandalism, protection against, 151, 156

VanDerLeeden, Tony, 82, 88

Velasco, Ralph, *20*, 21-38

Venezuela, 234-42

Vera Cruz, Mexico, 56

Verdi Cave. *See* Frio Cave

Virginia, 130-41, 178

Wales, Joseph H., 94

War: Civil, 108, 132, 134, *135*, 136, *137*, 209-10, 292; Mexican, 34; of 1812, 106, 121-22, 134, 209; Revolutionary, 134

Warner, Robert, 298

Washington, George, 130, 132

Water: hot spring, 90-104; movement underground, 5-8, 144; rise after storm, 108, 110, 283; supply, 4, 246, 260, 261; warm, 256. *See also* Ground water

Waterbury, Conn., 293

Watertown, Conn., 298

Weasel(s), 173-74, 270

Westchester County, N. Y., 290-303

Western Maryland Railway, 172-184, *177*

West Indies, 18

West Virginia, 3, 16, 17, 59-77, 133-40, 176, 181-82

Weyer, Bernard, 131

Weyer's Cave, Va., 131, 137

Whippoorwill Park, N.Y., 301

White, Jim, 149-50

White, Patrick J., 19-38, *20*, 212-13

Whitney, J. D., 194

Wills Creek, Md., 173

Wind Cave, Pa., 7

Wisconsin, 18

Wolf, 271

Woodbridge Cave, Conn., 300

Woodbury, Conn., 297-98

Wood rat. *See* Cave rat

Woods, Loren, 284

Wright, Harold Bell, 271, 278

Wyandotte Cave, Ind., 317-23, *319, 322*

Yokum Knob, W. Va., 138, Map, 140

Zemí. *See* Idol

Zigzag Salamander, 289

Zinc, 13, 18

Zone: ground-water, 5; phreatic (saturated), 8, 144-45, 147; twilight, 60, 240, *280*, 288